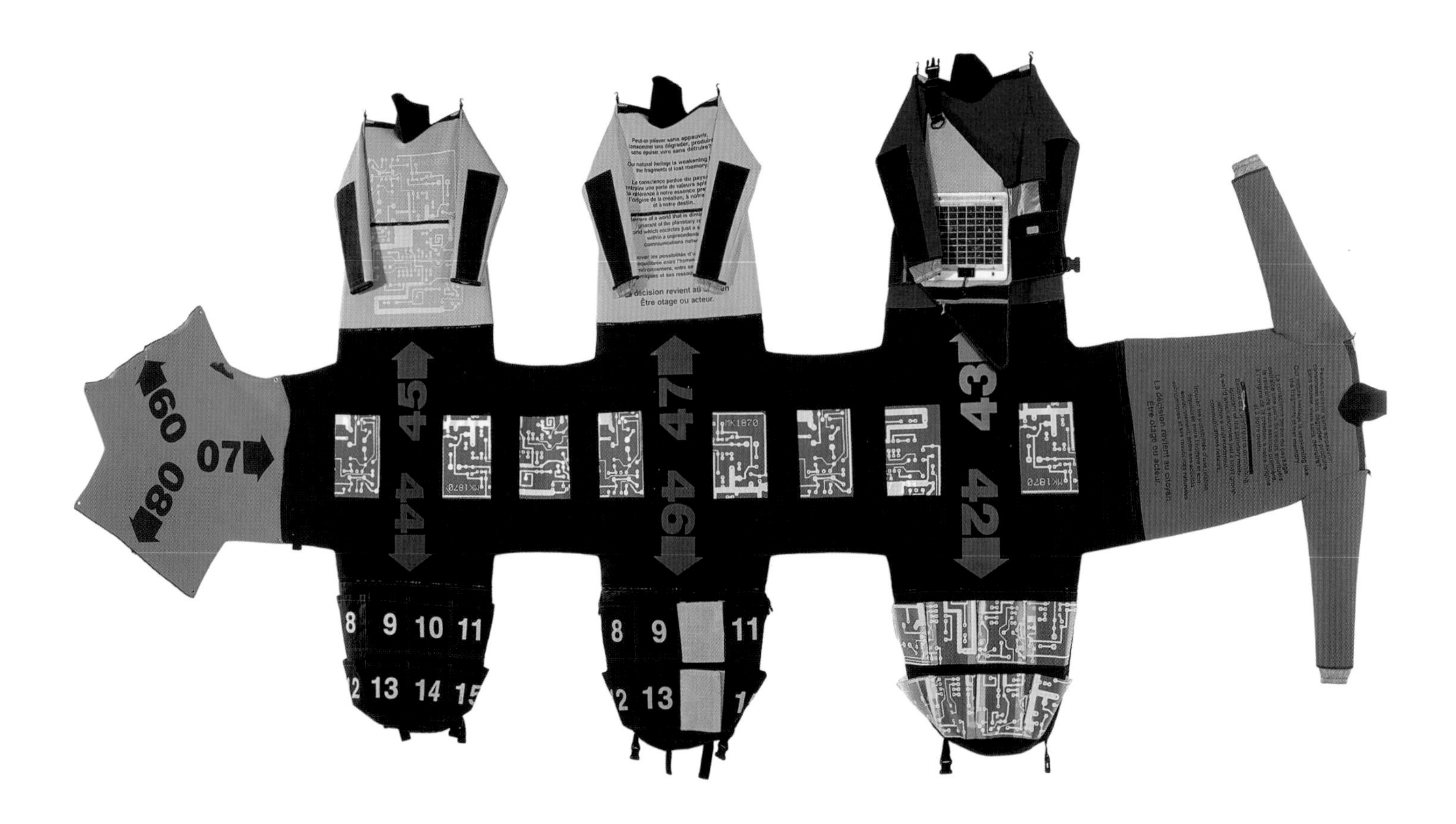

More Mobile

Portable Architecture for Today

Edited by Jennifer Siegal

Foreword by Jude Stewart
Introduction by William J. Mitchell

Princeton Architectural Press, New York

Dedication

Steven and Richard Siegal—the true global nomads
Gail Siegal—the center of our wheel

Published by
Princeton Architectural Press
37 East Seventh Street
New York, New York 10003

For a free catalog of books, call 1.800.722.6657.
Visit our website at www.papress.com.

Printed and bound in China
11 10 09 08 4 3 2 1 First edition

Project Editor: Clare Jacobson
Copy Editor: Dorothy Ball
Designer: Jan Haux

Special thanks to: Nettie Aljian, Sara Bader, Nicola Bednarek, Janet Behning, Becca Casbon, Carina Cha, Penny (Yuen Pik) Chu, Russell Fernandez, Pete Fitzpatrick, Wendy Fuller, Aileen Kwun, Nancy Eklund Later, Linda Lee, Aaron Lim, Laurie Manfra, Katharine Myers, Lauren Nelson Packard, Jennifer Thompson, Arnoud Verhaeghe, Paul Wagner, Joseph Weston, and Deb Wood of Princeton Architectural Press—Kevin C. Lippert, publisher

Library of Congress Cataloging-in-Publication Data
More mobile : portable architecture for today / edited by Jennifer Siegal ; foreword by Jude Stewart ; introduction by William J. Mitchell.
p. cm.
ISBN 978-1-56898-758-3 (alk. paper)
1. Buildings, Portable. I. Siegal, Jennifer, 1965-
NA8480.M67 2008
720—dc22
2008003927

Credits

This original version of the Jude Stewart foreword appeared in *DIRECTIONS* magazine, Spring/Summer 2007. Reprinted with permission.

All images courtesy of the contributors unless otherwise noted.
2: Studio-Orta
8: Courtesy Airstream
11: Office of Mobile Design
15: Michael Chia-Liang Lin, courtesy William Mitchell
16: Franco Vairani, courtesy William Mitchell
34–36, 39, 40R: Robert R. Roos
57–59: Andrew Maynard Architects and Buro North
65: Mauro Mattioli
71: Jean-François Jacq
121–123, 127: Danny Bright
130, 142–143: Undine Prohl
131, 133: Daniel Hennessy
141–143: Benny Chan/FotoWorks

Contents

6 Foreword by Jude Stewart
10 Preface by Jennifer Siegal
13 Acknowledgments
14 Introduction by William J. Mitchell
22 Studio-Orta
32 Dré Wapenaar
42 Andrea Zittel
52 Andrew Maynard
62 Andreas Vogler
72 Horden Cherry Lee Architects
82 N55
92 Atelier Bow-Wow
102 The Mark Fisher Studio
112 MMW
120 LOT-EK
128 Office of Mobile Design

Life on the Move

Jude Stewart

We're living more mobile lives than ever, driving change in design and architecture. Welcome to a roving new future.

Murmur the words "mobile lifestyle," and a slide-show of glamorous images strikes up on auto-pilot: lean, post-dot-com types in airports and rickshaws, riding the rails, swapping international backgrounds like theater sets, enhancing their natural powers via sleek bits of wireless technology. Mobility in the modern era, it seems, spells future-forward and affluent like nothing else.

Of course, every bright glint of "mobility" casts a shadow, too. Globetrotting and time- and space-bending gadgets have become remarkably, but not fully, democratized. As travel brings us closer to every corner of the world, it also strands us in blandly similar way stations, wasting record amounts of our time in transit. The global reach of business speeds imports to our doorstep, even as it homogenizes landscapes. If mobile technology frees us from the limits of place, letting us live and work virtually, it also can leave us feeling hollow, placeless, craving real people and locales. Perhaps most critically, our mobility depends on gobs of electricity (to power all those gadgets) and petroleum (to fuel all those planes, cars, and conveyances). Clearly mobility has kinks to work out as we march into the future.

Where does architecture—arguably the most rooted branch of design—fit into this moveable feast? In the fast-growing category of "mobile architecture," buildings pull up stakes and go places; factories assemble custom housing and whisk it to ready-made sites; materials work smarter and harder toward sustainability; temporary structures respond more readily to our desire for mobility, from disaster aid to ad hoc celebrations; and our collective sense of place, self, and society shifts. Welcome to the new, roving landscape of design.

The Factory-Built Future

Today's mobile housing offers more than the thrill of the open road. As real estate values and the likelihood of relocation creep ever upward, it's a bonus to be able to take your custom-built, energy-efficient home with you to a new city—or resell it, eBay-style, in a market that extends well beyond your neighborhood.

Mobile-home manufacturers like Canada's Sustain have taken the first giant steps forward in the category, putting a decidedly more eco-friendly spin on traditional mobile housing. Sustain miniHomes convert from wheeled trailer to rooted home, powered by solar panels, a wind turbine, and propane. Hook up your freshwater and wastewater tanks to a municipal source, and you're literally home free.

Sustain's genius lies in its factory manufacture, a larger-scale architectural trend made possible—and sensible—by increased mobility. Factory-built prefab lets architects "think about buildings like a product designer does," says Jennifer Siegal, principal and founder of Office of Mobile Design, a progressive architecture studio in California. Siegal's firm designs moveable buildings as well as prefab steel structures, built in factories and shipped to the building site, with remarkably quick turnaround. She considers traditional homebuilding inefficient, more like handicraft than a streamlined production process. "Would Audi's engineers cobble together their new model in someone's backyard?" she quips. "On a typical construction site [in the United States], 30 percent of materials get tossed in a dumpster, whereas in a factory 99 percent of those materials get used or recycled into a new project. Building in a controlled environment had benefits I hadn't even considered when I got started." Factory-built homes centralize materials, processes, and tools from all over the world, speeding a high-quality, sustainable home to a site—a home that can often be relocated later. Mobility is writ large in every step of the movement.

Efficiencies add up even faster when architects open their minds to a broader palette of construction materials. "A material palette that's stronger, smarter, and more lightweight hasn't really trickled down to the building industry," Siegal remarks. "Spacecrafts and even bicycles have made more advances [than construction]." She looks forward to bioengineered materials that replicate the structural strength of spiderwebs; in the meantime she opts for materials like the plywood alternative Kirei, a recycled waste product of Japanese sorghum grass, and non-VOC paints.

Recycled materials can even provide the design hook. Architect Adam Kalkin takes advantage of the current oversupply of shipping containers in the United States, a result of the soaring trade deficit. His Quik House building kit uses the containers to build a full-size home in a single day for US$125-$165 per square foot.

Ad Hoc Architecture

Better materials pose a fascinating paradox: their strength and resilience ensure longer-lasting "permanent" buildings, but those same qualities make them endlessly recyclable into new, temporary structures. These new buildings respond more flexibly to our changing mobility, springing up or moving to us as the need arises. Since 1989 Japan's Shigeru Ban has built temporary theaters, churches, even emergency housing from paper tubes treated with paraffin and glue and rooted in concrete or steel bases. Exposure to wind and ultraviolet rays actually increases the tubes' compressive strength, allowing nearly all the materials to be disassembled and reused. Similarly, UK-based Inflate creates reusable, inflatable structures for temporary use indoors, from its Office in a Bucket pods to event kiosks and retail environments.

Even furniture designers are responding in kind, with designs that respond more swiftly to the changing needs (or floor plans) of their owners. Anders Englund, design manager and co-owner of the Swedish furniture firm OFFECCT, puts it this way: "Good modern furniture design should satisfy the 'moment,' meaning it can be very permanent in one way, but you can use it in different ways for the situation." OFFECCT's Forest Room Dividers and the Cloud, an inflatable meeting room, change spaces dramatically with minimal setup. The firm also offers a line of sound absorbers that block noise while doubling as mobile room dividers. Other designs, like the Flower Stool or Woob Chair, use ultra-lightweight foam to make rearranging conversation spaces a snap. For Englund, living and working spaces work better when uncluttered and mobile: "I welcome concepts with less furniture that give me more alternatives and space to use them," he remarks.

Mobility Travels the World

The travel industry is perhaps the clearest bellwether for where mobility trends may be headed next. Glen Hiemstra, founder of Futurist.com and author of *Turning the Future Into Revenue*, sees big-time growth in travel for two reasons: retired, affluent baby boomers and the so-called digital-native generation born after 1981, for whom the internet, cell phones, and frequent travel have always existed. "It's like a force of nature, this wanting to see more," he notes. Hiemstra predicts virtual living will only sharpen our keenness for unique places when we travel: even with galloping advances in video conferencing and greater transparency via the web into all corners of the globe, Hiemstra says, "the need for experience will not diminish."

The Airstream Company, founded by Wally Byam, 1935

While mobile design has perhaps its strongest foothold in affluent places like Europe, the United States, and Japan, it is trickling down surprisingly quickly to the developing world, too. American futurist Howard Rheingold calls the three billion cell phones worldwide "the poor man's connection to the web." He continues: "The developing world uses cell phones to get the info they need; they may not think of it as internet access. Fishermen can find out where the good catches are; farmers can check commodity prices; subsistence workers know when there's day labor in a neighboring town." John Urry, sociology professor and director of the Centre for Mobilities Research in Britain's Lancaster University, affirms that mobility reaches even those who can't afford to travel constantly: "In poorer places there's still plenty of contact with mobility as hosts receiving visitors from afar." He also notes that poorer world travelers like students, au pairs, and casual workers are forming a growing, more democratic mobile class.

Mobile architecture can dovetail with affordable housing and other pressing third-world issues. South African architect Eric Bigot founded Zenkaya as a dual-action social project, providing first manufacturing jobs to locals and eventually better affordable housing. "I'd like to manufacture Zenkaya at an affordable cost [here], using and improving South African labor, and then finish and assemble them on-site in the States for that market," Bigot explains. If costs drop in wealthier places, he avers, it could jump-start prefab housing at all price points globally.

Home Is Where You Are

Perhaps mobile design's richest irony is this: it deals as much with staying put as with movement. Futurists predict mobile technologies like web-enabled video screens in our eyeglasses, supercharged video conferencing, even wireless devices embedded in our bodies. All achieve the triumph of self over the limits of place—without moving one bit. Constancy and rootedness in the midst of travel and change stand for something after all.

Melinda Stokes, an Australian clothing designer based in Berlin, creates clothes for travelers with this sense of constancy in mind. Her clothes stealthily incorporate a myriad of pockets and ergonomic fabrics, keeping the rough-and-ready traveler unfussed and stylish at all times and places. She sums up the paradox of mobile life very plainly: "When we travel, we don't actually transform. We're still ourselves, with the style we had when we left. It's quite a strong thing, clothing. There's a kind of protection about it; it gives you a particular sense of self." No matter where you're headed next, you'd hardly want to leave that behind.

Generation Mobile

Jennifer Siegal

The work assembled in *More Mobile* looks to a new generation of architects, artists, and designers from around the world that are in sync with new nomadism. They are a growing grassroots, do-it-yourself movement, impatient for change and looking for ways to inject the personal into the social, achieving a more fluid—and ultimately more authentic—experience of our contemporary cities and our changing culture.

Since the publication of my book *Mobile* in 2002, the world of portable architecture has changed dramatically. Conceptual projects have now been fully realized. Smaller, lighter, and more compact dwellings are sold online in countries including the United States, Australia, Norway, and Japan. "Smart materials," whose physical properties or outward appearance are designed to change, have created a new lexicon for these mobile building forms. In addition, an ever-increasing flow of digital information facilitates the immediate exchange of fresh ideas and expands our desire to explore the unknown.

More Mobile examines ideas that are not traditionally found in the pedagogy of architectural education. Today, however, the topic is taught in universities around the world where students and their professors are seeking out new design territories and modes of production. The book introduces some designers new to mobile design and re-presents some featured in *Mobile* whose work has transformed in the intervening years. The notorious rock-and-roll stage designer Mark Fisher expanded his repertoire with the creation of the Rolling Stones' 2005–7 *Bigger Bang* tour, the design of which is inspired by an operatic environment set in the twenty-first century. LOT-EK's theories are put into practice in its branded container stores for UNIQLO, which bring apparel to the closets of the consumer. New additions come from artists such as Studio-Orta, Dré Wapenaar, and Andrea Zittel, whose work can be political, theatrical, fashionable, functional, and wearable. Atelier Bow-Wow, another of the new contributing architecture firms, places mobile agitprop structures in urban settings as a provocation between art and architecture.

The future of mobile architecture is unfolding rapidly. As our buildings become more portable and adaptable, they become more useful. Before long we will shed the bulk and excess of static environments as we look to Generation Mobile and its long-term solutions for the uprooting of today's built structures.

Sheep Rock
Courthouse
Towers
Park Avenue
4982
RIVER
4510
5197

Acknowledgments

Clare Jacobson—the best and brightest mobile editor

Office of Mobile Design—the fluid members: Kelly Bair, Elmer Barco, Tim Barnard, Carina Bien-Wilner, Tim Bonefeld, Barrett Cooke, Stephanie Cutler, Saul Diaz, Joel Escalante, Matthew Fellows, Matthew Garcia, Gerardo Gazia, Wendy Gilmartin, Brian Hajjar, Larry Harris, Armando Hernandez, Barbara Huang, Lori Jay, Anne Marie Kaufman Perlov, Elizabeth Kelsey, Peter Klein, Deborah Lehman, Kevin Ly, Laura McAlpine, Ashley Moore, Thao Nguyen, June Okada, Jason Panneton, Damian Petrescu, Jon Racek, Christine Reed, Ariana Rinderknecht, Greg Roth, Lena Schacherer, Aaron Schump, Sara Schuster, Naoto Sekiguchi, Jacob Simanowitz, Jenna Sommers, Mark Stankard, Andrew Todd, Lorna Turner, Marine Vanyan, Arona Witte, Michael Zahn

A House Is a Robot for Living In

William J. Mitchell

In the digital network era, all machines aspire to the condition of robots.

From a strictly mechanical perspective, cities consist of fixed infrastructure combined with moving parts. The moving parts provide responsiveness, flexibility, and adaptability, but they also consume more energy than the stuff that just sits there. One of the most fundamental of design dilemmas, then, is whether it's worth putting a function into a moving part and continually expending the energy that this takes, or whether it's better to go for the economy of keeping it static. Design innovations—like many presented in this book—often propose putting normally static elements into motion or, conversely, getting some normally moving elements to settle down in one place.

Scale and Movement

Some potentially moving elements are huge, like icebergs, and some are small, like ice cubes. Some have few internal degrees of freedom and move as rigid wholes, while others, like flocks of birds, have many internal degrees of freedom, can change their overall shapes, and can move around as coordinated collections of individual parts. Internally connected things with articulated movements, like robot arms, are somewhere in between. Some movements are random and create only fleeting patterns, as with the Brownian motion of really tiny things, but others are regular and repetitive, creating persistent temporal and spatial patterns such as whirlpools. Some of these patterns have only statistical regularity, as with waves on the beach, while others, like those of a pendulum, are closely predictable.

Since force equals mass times acceleration and large forces are hard to achieve, it's rare for huge, rigid things to move around a lot. Nuclear aircraft carriers are an exception. In this case, the pressing mobility demands of warfare (usually, in practice, those of global intimidation) justify a massive onboard power plant and extraordinary expenditures of energy. At the other end of the scale, motes of dust in a beam of light have so little inertia that incident energy constantly bounces them about.

This ironclad logic produces a world in which big, heavy, slow-moving things with low-frequency patterns form a background for the faster movements and higher-frequency patterns of smaller things. When something violates this logic, like the lightly dancing hippopotamus in *Fantasia*, it grabs our attention.

Moving Parts

If you *do* want a big, heavy thing to move—especially to move quickly, it's usually best to move it in parts. A nomad camp isn't like an aircraft carrier. It adds up to a huge thing, but it consists of many independent pieces that are small enough to be picked up by people and put on horses. This doesn't mean that its pattern can't persist over time, since the pieces can then be put down again, in the same spatial relationships, in a new place. It's a much more convincing example of a walking city than that proposed by '60s pop theoreticians and designers Archigram.

In systems of rigid moving parts—like most mechanical systems and buildings—parts are related to one another in just four basic ways: some are rigidly connected; some are constrained to slide past each other; some are constrained to rotate around points or axes; and some, like bees in a bottle, move in unconstrained ways. Traditional drafting practices and constructions are specialized reflections of this: stencils hold graphic elements in rigid relationships; parallel bars slide vertically, and drafting pens slide horizontally against them; compasses pivot; and a freehand pencil is unconstrained.

In most buildings, the connections are rigid. Desk drawers, sliding doors and windows, elevators, escalators, and retractable roofs slide relative to the fixed infrastructure formed by rigidly connected, heavy construction components such as floor slabs. Similarly, swinging doors and sash windows pivot on rigid connections to this infrastructure. At low speed and frequency, items of furniture move around inside —mostly along the floor planes—in unconstrained ways. The human inhabitants are even more mobile, and occasional intruding birds and insects take unconstrained movement to higher speeds and the third dimension.

Electric, folding RoboScooter by the Smart Cities group at the MIT Media Lab in collaboration with ITRI and Sanyang Motors (SYM). Complex articulated motion.

Pivoting motion also provides the basis for legs and wheels—the essential components of ground transportation systems. Animals evolved legs to move around, mostly because legs work well with the rough, unpredictable terrain found in nature. Vehicles mostly have wheels, since these are far more efficient than legs where major investment has been made in hard, flat, generally horizontal surfaces such as paved roads and parking lots. Mountain villages that depend upon donkeys or camels have urban forms adapted to low-speed leg motion, while modern cities are much more adapted to high-speed wheel motion. It's possible that running, jumping robots and electronically manipulated exoskeletons, like those developed by Hugh Herr at the MIT Media Lab, might find modern urban transportation roles. But I wouldn't bet on this in the short term.

As the ingenious have long known, simple sliding and pivoting movements can be combined to create more complex articulations. The wheels of a railway carriage rotate, for example, while the passenger compartment translates. If you enumerate possible combinations of constrained motions, as the great engineer Franz Reuleaux did in the nineteenth century, you produce a catalog of kinematic mechanisms. These serve as archetypes. The catalog that Reuleaux published is the equivalent, for architecture in motion, of Jean-Nicolas-Louis Durand's earlier index of possible plans and elevations for neoclassical buildings.

Often, complex systems of moving parts branch into hierarchies of subsystems, and these are wrapped in covers that conceal what's going on inside. There's a disk drive whirring at high speed inside your computer, for example. And most of the complex choreography of a gasoline-powered automobile is concealed inside the smooth, rigid, sheet-metal body. Many designers devote themselves to styling and putting on these wrappers.

The alternative approach, of course, is to take the wrappers off. Motorcycle designers mostly favor this; they want us to feel the excitement of lots of parts rapidly and noisily in motion. Gorgeous red Ferraris make themselves even more seductive by exposing their rear engines, with all their beautifully crafted moving parts, beneath transparent engine covers. At a larger scale, architects sometimes like to expose moving elevators (alternatively wrapped in opaque shafts).

Nonrigid Movement

The expansion of a balloon, unlike a translation or rotation, isn't a rigid motion. It's an elastic transformation—one of scaling to a larger size. If you twist a balloon in your hands or create a balloon animal from it, you create more complex, differential deformations of the original shape.

At the level of drafting, splines—either old-fashioned wood or metal ones that are bent into the shape you want, or the mathematical, parametric ones provided by CAD systems—embody the idea of elastic deformation. Much like the operation of twisting sheets of paper to create curved-surface architectural models, crumpling sheets of paper extends the idea to nonelastic deformations.

Electric, robot-wheeled City Car by the Smart Cities group at the MIT Media Lab.

Modular assembly, folding, and omnidirectional motion.

In animal bodies, elements other than those of the skeleton are mostly deformable, and their articulated motions are complex combinations of translations, rotations, and deformations. Animal skins take various forms to accommodate this. They can be hard carapaces with openings for the softer elements, as with tortoises and mollusks. They can be systems of sliding scales, as with snakes. Or they can be smooth and elastic, as with humans.

Nonrigid systems frequently have elastic actuators, such as rubber bands, living muscles, or "air muscles" such as those produced by Festo. In animals these activate anatomies consisting otherwise of rigid compression and bending members (the skeleton); tension members (tendons); elastic tissue (flesh); and an outer skin. The effects of these sorts of systems moving fluidly under elastic membranes are endlessly varied and subtle. They can be sexy. And dancers know how to deploy them. That's why, of course, dancers tend to show lots of skin.

Such effects are harder to produce with artificial materials and actuators, especially at building scale. But inflated structures with elastic envelopes, cables as tendons, and Festo muscles or the like as actuators do provide some interesting and as yet little-explored possibilities. Actuated tensegrity structures, with or without skins, can move in lifelike ways. Chain mail has been around for a while. And more possibilities are now emerging from technologies for weaving high-performance materials, such as metal and carbon fibers, into structural fabrics—perhaps with embedded actuators.

Since Adam and Eve got tossed out of Eden, clothing has added another layer to the movements of people, animals, and animal-like artificial structures. Cloth has its own characteristic dynamics and patterns, with its capacity to fold, drape, stretch, billow, veil, cling, and zip up in a vast variety of ways. This has long been the specialized domain of fashion designers, and buildings have traditionally been clothed—partially at least—with drapes, awnings, and rugs. Christo's burka-like projects have taken the idea of clothing a building still further, and they might become even more engaging if the underlying structures can be made to move around like bodies under a bedsheet.

Lifelike architectural skins and clothing can, of course, be sectional and multilayered. This opens up the further choreographic and ritual possibilities of wrapping and unwrapping, peeling, winding and unwinding, enrobing and disrobing, billowing up like Marilyn Monroe's skirt, running it up the flagpole, furling and unfurling, cross-dressing, wardrobe malfunctions, and striptease.

Discontinuous Transformations

So far we have considered continuous translation and rotation of building parts and continuous deformations. Mathematically, there is also the possibility of discontinuous transformation, in which a part of some object is simply replaced by something else. The "replace" operation of Microsoft Word, for example, substitutes one string of characters for another.

In the material world of actual buildings, it occurs when a plumber takes out a fixture and replaces it with another one. In larger-scale renovations, bricks and stones and timbers might be replaced with new ones. At a larger scale still, an owner might replace an entire roof or demolish and replace a complete wing of a building.

This is simplest when the replacement part has exactly the same form and material as its predecessor. In other words, there is a transformation of substance but not of form. Over time, old buildings may have their materiality completely transformed in this way, until there is no trace left of the original materials. Most of the geometric and physical properties of the building are preserved under this sort of transformation, though, so we generally have no hesitation in saying that it remains the "same" building.

A more complex case occurs when the substituted part has the same form but a different type of material, as when modern plaster replaces ancient stone in an archaeological reconstruction. Fewer of the building's properties are preserved in this case, which can lead to arguments about authenticity.

Still fewer properties are preserved when replacement parts maintain an interface to surrounding parts but have different forms and materials. Kit-of-parts approaches to design and construction and online product configurators—including those that are starting to appear for dwellings—follow this logic. Here we tend to think of the results of transformations as new works of architecture—variations on the theme established by the original—rather than as continuations of the existence of the same building. It all depends, though, on how much reconfiguration occurs and on where we choose to draw the arbitrary line between "same" and "different."

In summary, architecture in motion is a game of transforming buildings and parts of buildings. The transformations that are applied may be translations and rotations, which preserve size, shape, and material properties. They may be developed to continuous distortions, which allow sizes and shapes to vary as well. And finally they may be extended to various kinds of discontinuous transformations.

Control Systems

The great contribution of the industrial era to architecture in motion—from Reuleaux to Archigram, was the technology of mechanisms triggered either by the human hand or by steam, that is, the combustion of hydrocarbons, or electric power. That of the digital network era is the microprocessor.

Animals have nervous systems—and, in more advanced instances, brains—to control the operation of their muscles. Similarly, today's machines—including the actuated parts of buildings—have digital networking and embedded microprocessors. These are crucial. In a typical modern automobile, the twenty-first-century parts—the electrical and electronic systems—cost much more than the nineteenth-century parts—the engine and power train.

Traditionally we have thought of machines with intelligence as robots, but with the ubiquitous embedding of networking and microprocessors in our everyday environment, most mechanical and electrical devices are evolving into robots. Vehicles are becoming wheeled robots, and buildings are becoming rooted robots composed of hierarchies of robotic subsystems—some attached, some free-floating. In the digital network era, all machines aspire to the robot condition. And a house is a robot for living in.

One of the first architectural robots was the automatic front door of the Massachusetts Institute of Technology at 77 Massachusetts Avenue. It has been there for many decades, and the great artificial intelligence pioneer Marvin Minsky once told me that it was the thing that convinced him to come to MIT as an undergraduate. It has a sensor to detect the approach of pedestrians, a simple control system, and an actuator that opens and closes the door in response. Essentially, it has reflexes that respond to relevant changes in its environment.

A more recent pioneering example is Jean Nouvel's Institut du Monde Arabe (1987) in Paris, which has actuated adjustable sunscreen apertures with photoelectric sensors. It suffers from the Achilles' heel of most such projects—a large number of mechanically prompted components that don't always work reliably. Furthermore, operation of the apertures' internal mechanisms consumes energy, and this partially negates the energy savings resulting from the responsive sun screening. Still, it was a landmark project.

In most significant buildings today, the HVAC systems provide robotic respiration. Their sensing and control systems are now digital, and their control software is increasingly sophisticated. Similarly, water supply and wastewater removal systems provide robotic alimentary tracts. Water filtering and purification, heating and chilling, distribution (drip irrigation of gardens, for example), wastewater treatment, and recycling processes have developed a great deal in recent years and are beginning to benefit from digital sensing and control. In general, buildings are evolving robotic physiologies.

Robotic physiological systems are typically unobtrusive, but they can be put on display when the occasion calls for it. Diller + Scofidio's Blur Building (2002), on Switzerland's Lake of Neuchâtel, with its 13,000 computer-controlled fog nozzles, makes responsive architecture out of moving mist. The Water Wall, developed by the MIT Media Lab's Smart Cities group for the Zaragoza Digital Mile and 2008 World Expo, uses digitally controlled solenoid valves to create programmable curtains of falling water. These can display images and text as patterns of water and air. (In effect, it's a one-bit-deep scrolling display.) More interestingly, it has sensors that can detect the approach of pedestrians and then open and close in response, like the Red Sea for Moses. Thus it creates dynamically located doors. If you want windows, you can throw a ball at it and a computer-generated one will automatically drop down to meet it at exactly the right spot.

Household appliances are evolving into robots as well. Vacuum cleaners, for example, were once dumb devices that you laboriously pushed around. But iRobot's Roomba, first released in 2002, is marketed as a "robotic floorvac." It's a small disc with infrared sensing that slides under furniture and runs around the floor under battery power. It detects dirt and sucks it up and automatically goes to wall plugs when it needs to recharge its batteries. It's nearly as smart as my dog. Watching it search for a plug, particularly if you thwart it with some obstacles, is a new form of slightly malicious high-tech amusement. If you're more ambitious, you can hack it to function as a mobile, light-seeking houseplant holder.

With the rapid development of digitally controlled LED technology, architectural lighting is leaving hot filaments, gas discharges, and old-fashioned light switches behind, and it's going solid-state and robotic. Where lighting systems once had small numbers of large incandescent or fluorescent fixtures, buildings of the near future will have large numbers of small LED fixtures distributed across their surfaces. Each of these will be addressable, like the pixels on a computer screen. Indeed, the difference between computer screens and architectural surfaces will begin to fade away. There will simply be different densities and patterns of light-emitting points as appropriate for different contexts and purposes. These systems will, of course, be connected to photoelectric and other sensors and will be run by microprocessors and control software.

The Free Pixels project from the Smart Cities group explores this idea. In a Free Pixels system, each lighting element is autonomous. It consists of photovoltaic cells for power harvesting, batteries for storage, LEDs for light output, and some electronics for control and wireless networking. You can hang collections of them up anywhere in regular or random patterns; you can rearrange their patterning and they automatically configure themselves as ad hoc wireless networks. You can then program them, just like very large computer screens. If you want them to be dynamically interactive, just add appropriate sensors and control logic.

Until recently, the only practical way to preserve spatial relationships among elements of an architectural composition was through joints and mechanical linkages—welds, tethers, hinges, sliders, and so on. Now, by means of embedded sensors, intelligence, and actuators, it is increasingly possible to preserve relationships through dynamic control. A work of architecture might become more like a flock of geese flying in formation or a military marching band. Each autonomously moving element is responsible for preserving the correct relationships with its neighbors. Architectural composition becomes robot choreography, or the planning of plays in robot team sports.

Conclusion: Objects and Processes

In the end, the difference between static architecture and architecture in motion is all a matter of the perspective you take. One of the great insights of modern physics is that you can look at physical things as both objects and processes. A whirlpool, for example, is clearly a fluid-flow process—an affair of water molecules in orderly motion. But it has a definite spatial location that can be marked on a map, and it persists in that location, so it is equally reasonable to regard it as an object. If it persists long enough, then it will be recognized as a local geographic feature and get a name, despite the fact that Heraclitus was right: you can't step into the same whirlpool twice. It's much the same with standing waves and with bands marching in formation. At a quantum mechanical level, light consists of electromagnetic processes—waves—or, when it's more convenient to look at it this way, of discrete moving objects—photons.

It is the same with biological systems. As Carl Woese formulated the point in his well-known paper "A New Biology for a New Century," living things are dynamic patterns of organization in the streams of energy and materials that pass through them. If we take snapshots of these patterns, they look like objects. If we consider their inputs and outputs over time, they look like processes—with patterns of organization that continually form and reform themselves.

At an architectural scale, the difference between objects and processes is also a matter of time frame. Walk up to a building, and it looks like an object. Take a longer view of its fabrication, assembly, occupancy, repair and transformation, and eventual disassembly and removal, and it looks more like one of Woese's dynamic patterns of organization in streams of materials and energy. We perceive architecture in motion when some of the ongoing transformations of a building—enabled by readily moved or substituted parts, actuators, and control systems—are fast enough to capture our attention and engage our imaginations. Or, to formulate the point in a complementary way, when their dynamic patterns of organization form and reform themselves at a rapid enough rate to be perceptible.

Studio-Orta

Founded in Paris in 1991, Studio-Orta operates as a research and development studio for artworks and limited editions by contemporary artists Lucy Orta and Jorge Orta and as an administrative bureau for their exhibitions and commissions. The two artists work in partnership, sharing a common research directive, and also work independently on special projects. This trilogy requires a unique cast of players, including curators, designers, architects, engineers, musicians, artisans, fabricators, production assistants, and technicians.

Employing a range of techniques from sculpture, object making, couture, painting, printing, light projections and site-specific performances, and public interventions, the team investigates crucial themes of the world today: the community and the social link, dwelling and habitat, nomadism and mobility, sustainable development, ecology and recycling.

Among Studio-Orta's most important contributions to these social and environmental debates are Refuge Wear and Body Architecture, portable minimum habitats bridging architecture and dress; HortaRecycling, the food chain in global and local contexts; 70 x 7, the ritual of the meal and its role in the community; Nexus Architecture, alternative methods to recreate the social link; Life Nexus, the metaphor of the heart versus the biomedical ethics of organ donation; OrtaWater, the general scarcity of this vital natural resource, the problems arising from the pollution and corporate control affecting access to clean water for all.

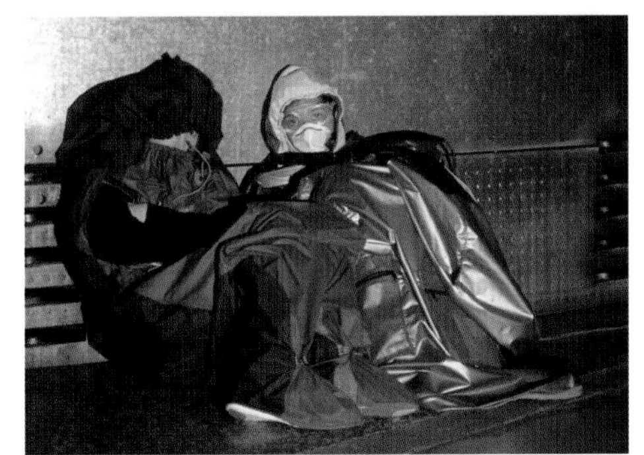
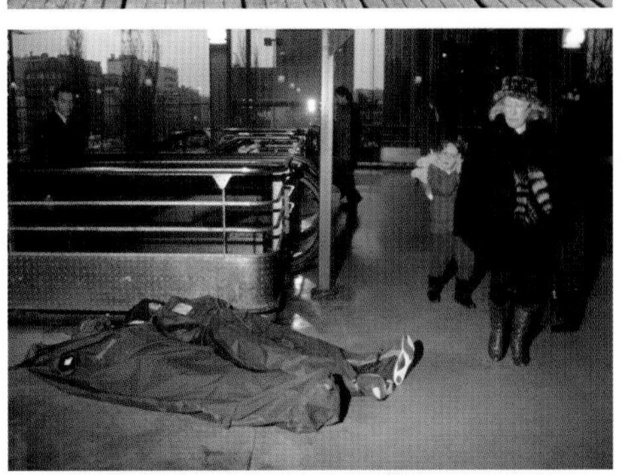

Dwelling

These Dwellings, commissioned by Angel Row Gallery, Now, and the Nottingham City Council, are prototypes for social spaces in the form of ambulating public sculptures that occupy city plazas, parks, shopping centers, museums, and housing estates. Each Dwelling comprises a compactable amorphous space frame structure and an inflatable patterned membrane, protruding from the bed of a converted military vehicle. Central to the artwork is the collaborative workshop component, which involves Lucy Orta working with local artists, facilitators, and community groups.

Each participant reflects on the notion of a social membrane through voluntary tactile workshops and investigates personal and collective spaces, mobile and connective. The consolidation of the personal creations originated from these research and development workshops forms the Dwelling "sketchbook." Further collaborations with engineers Atelier One enable the Dwelling to take on its final architectural form, a metaphor for the unification of culturally diverse groups into a collective whole.

Photo documentation, videos, sound recordings, and the participants' designs, scale modules, and technical information were collected at the end of the project and sent to the Studio-Orta team based in Paris to be analyzed and engineered into full-scale modules and segments. This was a highly involved process of translating abstract concepts and rudimentary sketches and paper models into a final cohesive 3-D form, matching colors, fabrics, and accessories to the technical specifications of each participant as closely as possible.

Once the fabrication at Studio-Orta was completed, all the participants' designs were archived in data drawers specially created for each iteration. The finished Dwelling was transported back to the community, to be assembled with the help of the community, where the social and structural layers of involvement were finally revealed.

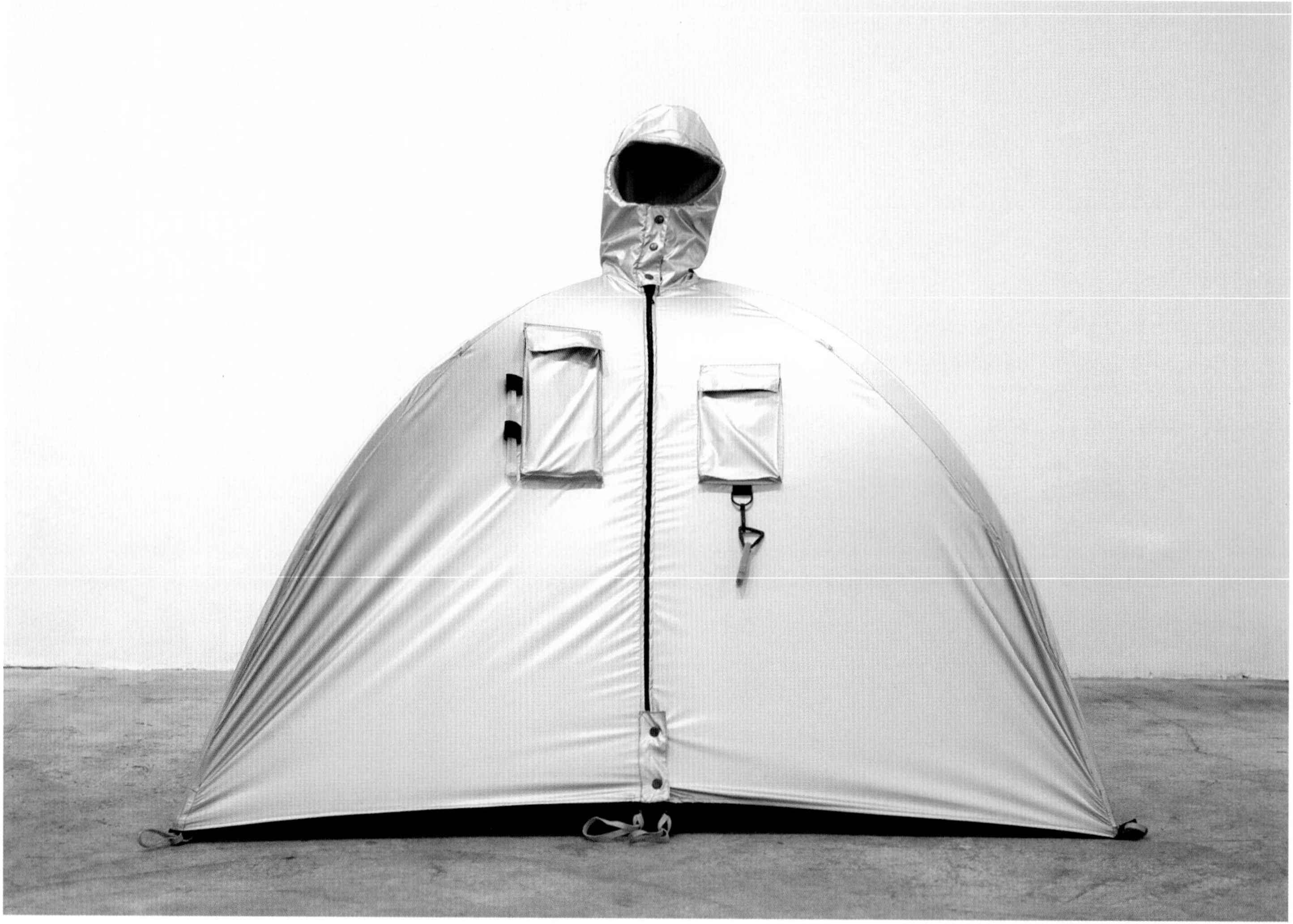

Nexus Architecture

Concept by Mark Sanders

Conceived as a process that forges the possibility of collective action and collaboration into a combined force, the various emanations of Studio-Orta's Nexus Architecture lead inward toward a renewed level of social activism. As a series of interdependent sculptures, they manifest themselves as the result of a personal decision of the artist to re-enter a community structure as an act of liberty, which investigates the power of solidarity through the symbolism of linkage.

Whether within the confines of the Cartier Foundation in Paris, or through the bringing together of members drawn from the Usindiso Women's Shelter at the Johannesburg Biennale, or from the citizens on the streets of our cities, Orta's Body Sculptures take on a social pertinence far beyond the parameters of a casual performance. Through a series of people dressed in outlandish, postatomic, even pre-apocalyptic overalls, we witness the birth of a new multifaceted, multi-limbed formation, numbering upwards of a hundred people joined together in a single human chain, sharing a common space.

Nexus Two: The Hanging Garments

Studio-Orta crafts personal wearable architectures that, once occupied and mobilized in city spaces, operate as miniaturized countersites. As architectures and performances, its works trace meandering itineraries and activate sporadic critical inversions that comment upon and contest taken-for-granted conditions of contemporary urban life: displacement, exile, poverty, homelessness. Its projects are never merely utopian, because they are both "mythic and real": they are realized, fabricated, and deliberately situated in the city but sustain a mythic aura because they resist the realist, propositional mode of the architect, urban designer, or planner. Suspended in this space between reality and myth, Orta equips and choreographs her urban subjects such that they stand witness to the troubles of the world. Orta's projects disturb our received picture of the world and solicit us to attend to it anew. Like Foucault's heterotopic spaces, they are catalysts for new attitudes to the city and the way we live in it.

HEART
LINK
NEXUS
REACT
PARTAGE

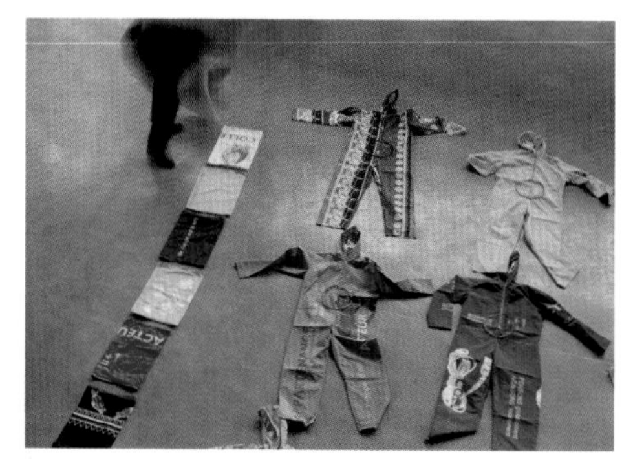

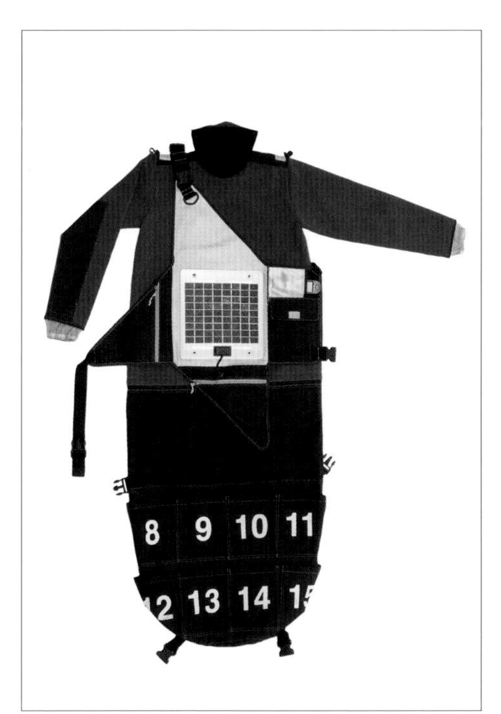
8 9 10 11
13 14

Dré Wapenaar

Dré Wapenaar works at the junction of art, architecture, and design as a choreographer of spatial, spiritual, and social encounters. His simple and light tent structures, sometimes described as social sculptures, are easily moveable and manipulatable by their specific users for very specific life events. Each sits lightly on the land (or on the museum-gallery floor), with a keen understanding of structural components, materials, and how light and the surrounding environment affects the mood inside and out. Wapenaar has built and orchestrated a tent camp in the shape of a ring of mushrooms; a newspaper kiosk secretly posing as a communications satellite; and a series of tents for life's most intimate moments, such as the Birthing Tent (de Baartent, 2003) and the Bivouac for the Dead (Dodenbivak, 2002). Occasionally Wapenaar's structures have a public function too, like his Tree Tents, designed as housing for a tree-saving activists in England, and his Newspaper Kiosk for the Rotterdam public library.

Born in the Netherlands in 1961, Wapenaar now heads the Studio Dré Wapenaar in Rotterdam. His most recent works are staging areas for our most fleeting moments of contemplation. The Pavilion of Emptiness, a portable, semitransparent, temple-like structure, allows the public to stop and reflect during its busy day. The Lumberjack's Pavilion is equipped with a hearth and chopping area for wood, complete with a hatchet and blocks of timber.

Wapenaar has participated in group shows like Stands and Tents at the Center for Visual Arts in Dordrecht (1999), Shine: Wishful Fantasies and Visions of the Future in Contemporary Art at the Museum Boijmans Van Beuningen in Rotterdam (2003), and Safe: Design Takes on Risk at MoMA (2005).

The Birthing Tent (de Baartent)

The Birthing Tent is, as the title indicates, a shelter for the ritual of childbirth and a tribute to new life. The fantastic, spherical structure guarantees a smooth transition from womb to external world in a place where the baby can be born beneath the stars.

The large bowl structure is equipped with benches for husbands, family, friends, and nurses to participate in the experience. Dré Wapenaar says of his work, "In my latest tents the theme of the loner versus the group, and/or the other, comes into a phase in which the interaction and the expectations that come with it will be directed even more. Here things deal more with emotional issues, with intimacy and distance."

The Birthing Tent also includes a bath where the mother can await her contractions, so that the transition from womb to outside world is a smooth one for the newborn baby. The roof of the tent can be opened so that the baby "can be born in safe surroundings beneath a starry sky," says Wapenaar. The Birthing Tent is mobile too; it folds up easily for one person to carry. Alternately, the tent could provide a home for those trespassing or camping out in a public space. It facilitates autonomy and, as Wapenaar has pointed out, perhaps no one except Buckminster Fuller has explored the possibilities of tents more than he has.

Newspaper Kiosk

The Newspaper Kiosk was deployed in the Rotterdam public library and in the atrium of The Hague's city hall, specifically intended to cause a reaction in these high-traffic public areas. It is precisely in such surroundings that the piece comes into its own, standing like a communications satellite returned to Earth from space. In airy defiance of the surrounding hustle and bustle, readers sit with their backs to one another on a circular wooden platform in the middle of the tent. This attitude allows them to immerse themselves in the newspapers placed on the counter in the outer circle. Passersby only see the readers' legs, resting on a circular bar inside. In this tent users seem to orbit around the interior of the kiosk, which carries the metaphor of a satellite for communication. The reading does take place in temporary isolation, but the shared seat can nonetheless make it an occasion for contact between readers.

Tree Tents

Tree Tents were originally designed for the Road Alert Group, a group of activists who fight against the excessive construction of highways through forests in Britain. During their protests, the Road Alert Group members cover themselves and hide (and occasionally live) in the trees so as to fight as long as possible against the incoming timber crews and their chain saws. Tree Tents were designed to provide a comfortable place for the activists to stay during their habitation of the forest, a stay that might prevent further forest destruction. The demonstrators nestle close to the treetop, swaying in the wind, out of reach of the arm of the law.

The form of the Tree Tents was derived from the simple physics of nature. Wapenaar envisioned a hanging pod that allowed its inhabitant to live within the tree—like a human birdhouse. When he hung a series of circular, sheathed platforms, tied by a rope to the side of a tree, as a conceptual test, the taut top and the bowed-out bottom of the dewdrop form came to be the driving geometry for the dangling pods.

As Wapenaar was designing the tents, a representative of the Road Alert Group convinced the designer to sell them his project drawings. The group finished the development of the tents, which were hugely successful for the group, both in terms of utility and as a formal identifier for the activists.

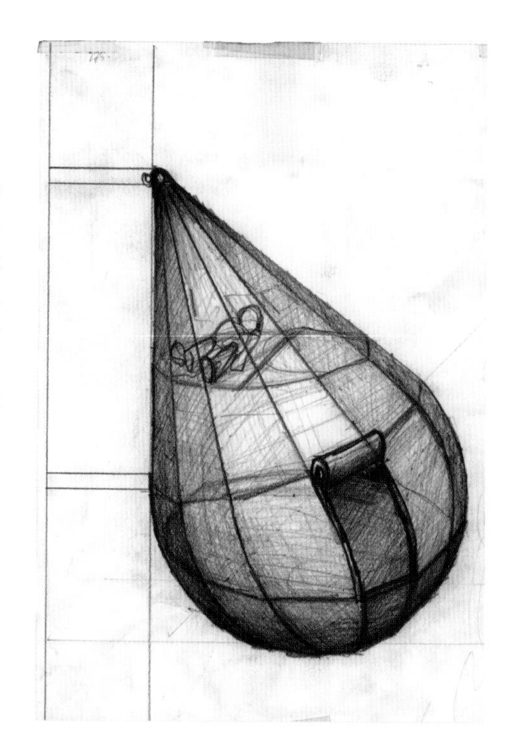

Andrea Zittel

Since the early 1990s, Andrea Zittel has used the arena of her day-to-day life to develop and test prototypes for living structures and situations. Using herself as a guinea pig, she often uses her own experiences to try to construct an understanding of the world at large. The experiments have at times been extreme—such as wearing a uniform for months on end, exploring limitations of living space, and living without measured time. However, one of the most important goals of this work is illuminating how we attribute significance to chosen structures or ways of life and how arbitrary any choice of structure can be. Without denying the personal significance of these decisions, Zittel's work tries to combine values such as "freedom," "security," "authorship," and "expertise," with an interest in how qualities, which we feel are totally concrete and rational, are often subjective, arbitrary, or invented.

Zittel was born in 1965 in Escondido, California. She received a BFA in painting and sculpture from San Diego State University (1988) and an MFA in sculpture from the Rhode Island School of Design (1990). In the early 1990s she established her practice in New York; one of her most visible projects there was A-Z East, a small row house in Brooklyn, which she turned into a showroom for her prototypes for living. Her work has also been included in group exhibitions at the Venice Biennale, Documenta X, Skulptur Projekte Münster, and in both the 1995 and 2004 Whitney Biennials. In 2008 she moved back to the West Coast, eventually settling in the High Desert region next to Joshua Tree National Park. She divides her time between A-Z West, located in Joshua Tree, California, and Los Angeles, where she teaches at the University of Southern California. She is a co-organizer of the High Desert Test Sites and is currently organizing two new projects: the A-Z Smockshop in Los Angeles and an as-yet-unnamed campground in the High Desert.

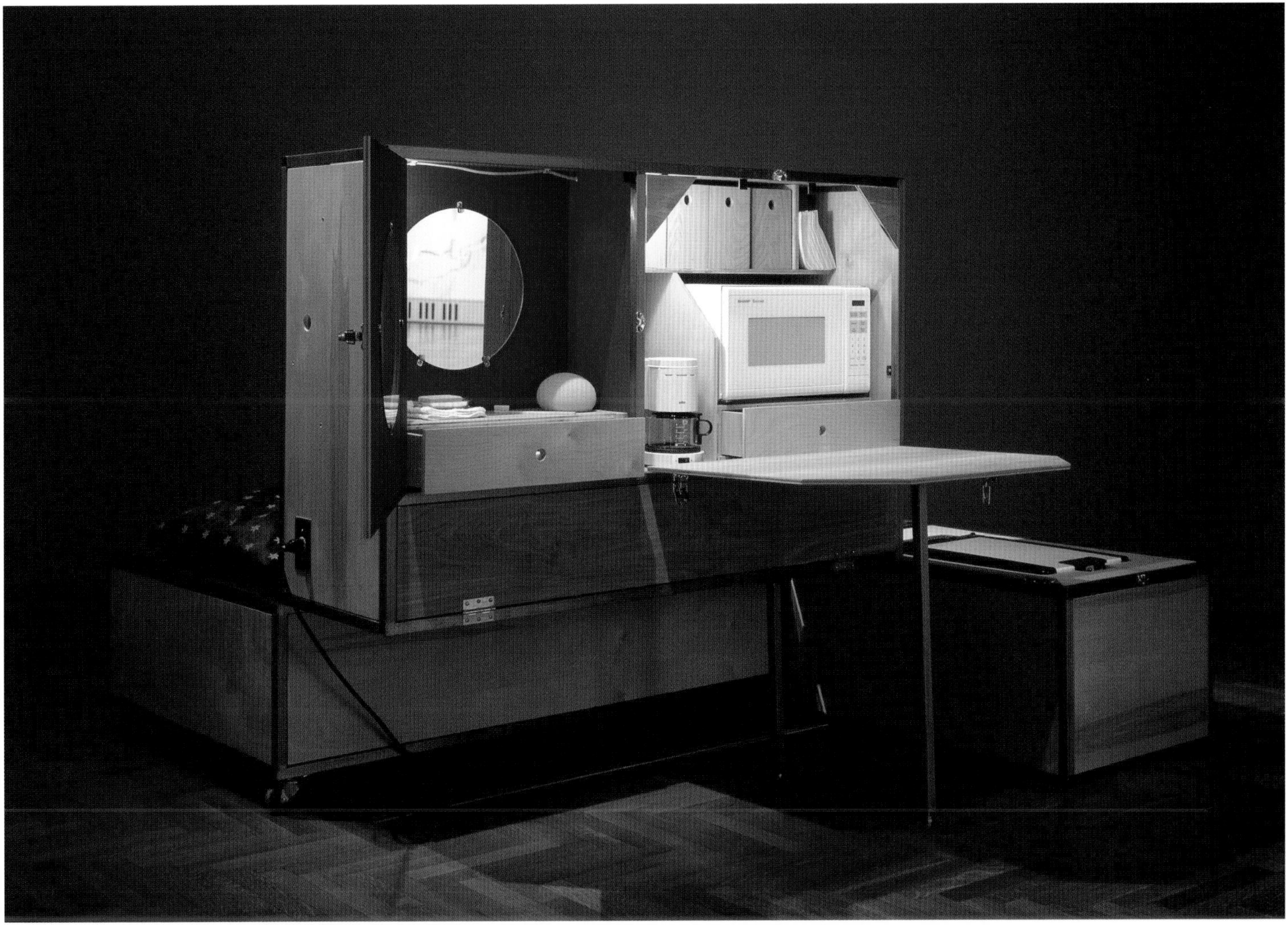

A-Z Cellular Compartment Units

The A-Z Cellular Compartment Units are a series of interconnecting boxlike chambers that transform the interior of a standard one- or two-room dwelling into a dense network of small rooms, facilitating a vast variety of functional and fanciful needs.

Although the Cellular Compartment Units serve as a functional habitat, the emphasis of the project was understanding the psychological implications of the space. The original Cellular Compartment Unit structure was fabricated as an off-site exhibition at the Ikon Gallery in Birmingham, UK. After it was built, Zittel lived in the structure (along with a few volunteers) for a month. The unit had a kitchen, three bedrooms, an office, a drawing room, an entry room, a dressing room, and a TV room. Although the space was tight, it proved comfortable, and up to four people could use a room at a time.

In addition to serving as an experiment in living space, the project took on a significant level of social commentary by pushing to the extreme an architectural format that most of us experience every day. This stacking and compacting of contemporary urban architecture contrasts with historical models of domestic dwelling, such as that in the Middle Ages, when a large household of both related and unrelated occupants would live together in a single hall eating, sleeping, and working.

As our contemporary lives and our buildings are divided into increasingly isolated compartments and categories, even the interiors of modest homes are divided by function into living rooms, dining rooms, family rooms, breakfast rooms, kitchens, laundries, and so on. This division by function also reflects a modern way of conducting our lives, by which we organize our activities sequentially by time codes and by task, much like the Taylorized division of labor implemented by the factory system and mass production.

As an increasingly dense network of rooms is installed in suburban mansions—gyms, hobby rooms, libraries, bars, and entertainment rooms—Zittel continues to imagine new programs and functions for the chambers in her future A-Z Compartment Units.

A-Z Wagon Station

The original pioneering spirit of the "frontier" considered autonomy and self-sufficiency prerequisites of personal freedom. At A-Z West Zittel continues to investigate how such perceptions of freedom have been re-adapted for contemporary living. She believes that, in this culture, personal liberation is more often achieved by individuals who manage to "slip between the cracks." Instead of building big ranches and permanent homesteads, today's independence seekers prefer small portable structures that evade the regulatory control of bureaucratic restrictions, such as building and safety codes. The A-Z Wagon Station reflects the qualities that create independence for the owner and user: compactness, adaptability, and transportability.

Two different associations connected to wagons inspired the design and concept of the A-Z Wagon Station. Similar to a covered wagon, the A-Z Wagon Station is intended to house possessions and provide a membrane against the elements. But the scale of the Wagon Station is actually derived from the dimensions of a standard-size station wagon, which provides the minimum space needed to create privacy and comfort for the occupant. Although the unit doesn't have wheels, the entire body of the wagon breaks down into five sections and can be transported to almost any location and reassembled by two people in only an hour or two.

Since their conception five years ago, sixteen of the eighteen original Wagon Stations have been customized. Most of these units are now situated in the desert backdrop of A-Z West, where they function as sleeping stations and simple camping shelters for Zittel's community of friends and collaborators.

STAY HOME!
Kerry Edwards
R
ROYA
6E54911

Andrea Zittel

Andrew Maynard

Recently named in *Wallpaper** magazine's Architects Directory, an "annual guide to the world's most innovative practices," Andrew Maynard's design practice is quickly becoming recognized as an emerging force on the architectural scene. Since Andrew Maynard Architects was established in late 2002, it has been recognized internationally by media, awards, and exhibitions for its unique body of built work and its experimental conceptual design polemics.

In 1996 Andrew Maynard received a bachelor's degree in environmental design from the University of Tasmania, and in 1998 he graduated with honors in a bachelor's in architecture from the same university. In 1998, with friend Stephen Mees, he won the Graphisoft International Design Award for the design of the Devil's Ballroom.

In 2000 Maynard won the Australia/New Zealand regional award and Grand Prize in the Asia Pacific Design Awards for the Design Pod.

Returning to Melbourne in 2002, he went into private practice, undertaking residential and commercial work as well as numerous design competitions. Maynard has exhibited at SMALL'03, an exhibition of emerging architectural practices, and been a guest speaker at PROCESS and at SOM'04, the Australasian Architecture Student conference. In 2005 he was a guest tutor at the fourth Australian Timber Design Workshop and spoke at the 2005 DesignEx International speaker series.

In September and October of 2004 he was an exhibitor at Art Directors Club Young Guns in New York. The Young Guns exhibition recognizes international designers thirty and under. Recently Maynard received a high commendation in the Australian Timber Design Awards for the Sproule residence and a special commendation in the VicUrban Affordable Housing competition for his Prefab House.

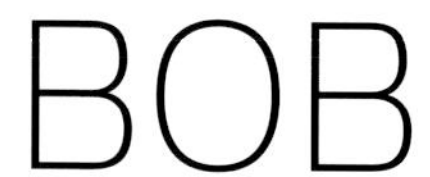

BOB

BOB is a hybrid home of the future, a mobile living tool for tomorrow's generation of nomadic wanderers. Somewhere between a tent, a house, and a Winnebago, BOB explores the relationship between the basic human requirements of travel and shelter.

In constructing BOB, the van's engine, gearbox, and drive train were relocated under the front seats, leaving the rest of the van for living in. Since all of BOB's technology and critical components are efficiently contained in the front quarter, the walls and roof are designed to be dynamic: they fold down and triple the effective floor space.

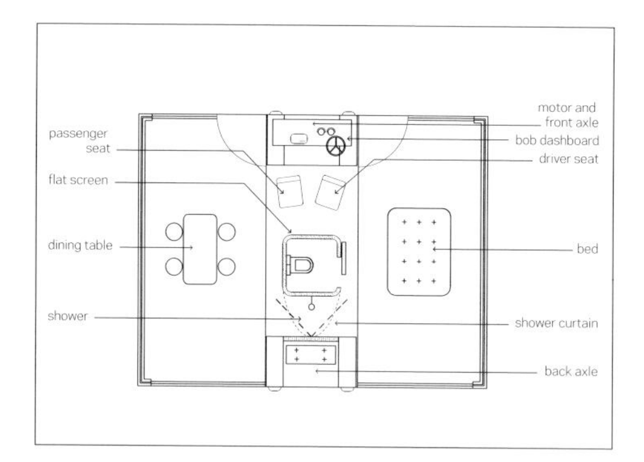

Unlike the dimensions of most motor homes, BOB's aren't dictated by a fixed width based on car lanes, and the internal elements don't need to be lined up one after the other with a tiny circulation path running through the middle. At the end of a hard day of driving, BOB spreads out and relaxes to create more square footage than his original footprint. Who said you couldn't have open-plan living in a mobile home?

BOB's cost depends on how many meters one desires, as BOB's price tag is based on the linear meter. Sort of like a stick of timber.

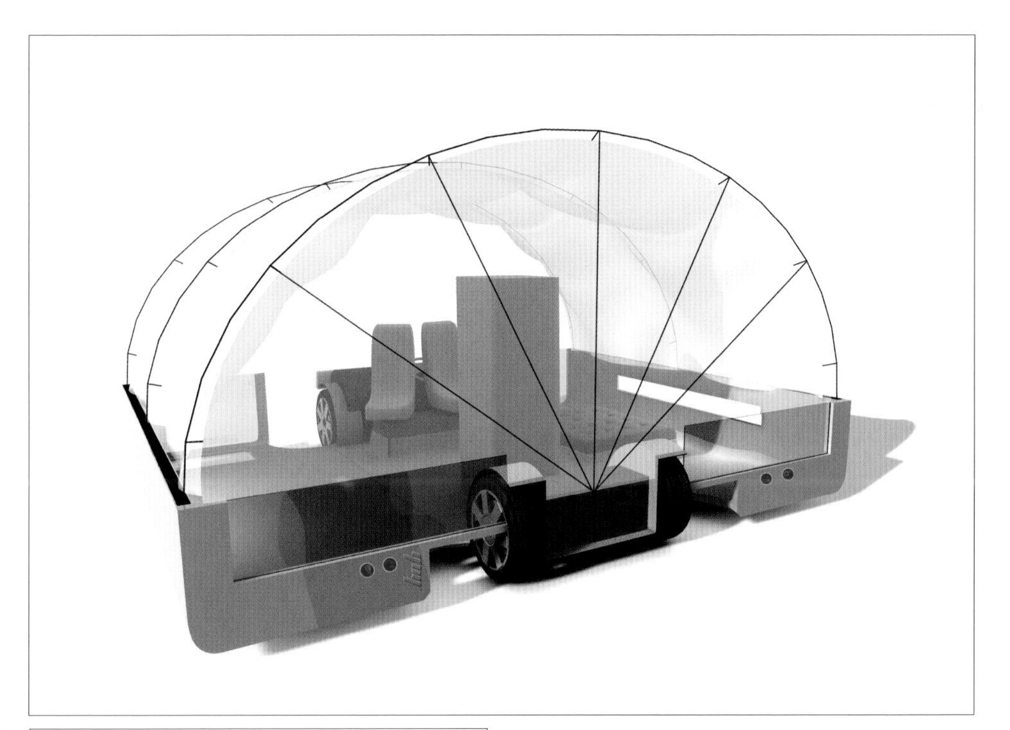

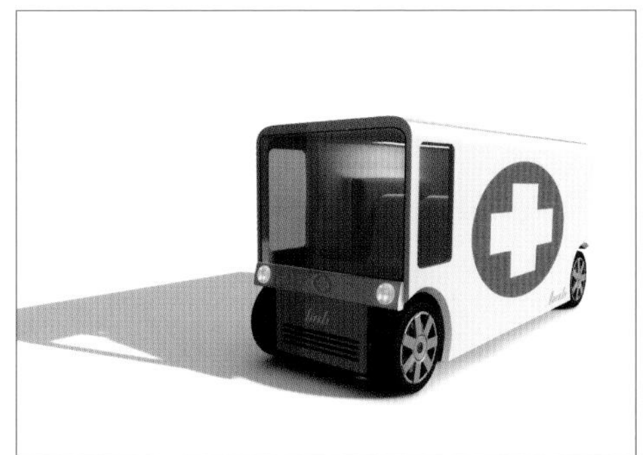

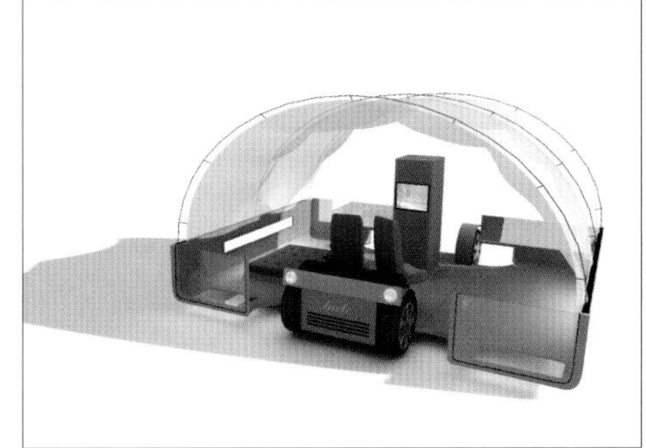

Corb V2.0

Not another container project...Containers provide terrible proportions for living in. Ever wanted to live in the penthouse? Want to get away from your annoying neighbor with the big stereo and bad music taste? Want to have a party without disturbing others? You want a different view every now and then? Corb V2.0 gives you the opportunity.

In *Towards a New Architecture* Le Corbusier wrote about the new epoch of housing he saw as intrinsic to the modern technological achievements of man. It was the machine that would make a better world. Through density, housing would not only be cheaper, but far better. "The problem of the house is a problem of the epoch. The equilibrium of society today depends upon it. Architecture has for its first duty, in this period of renewal, that of bringing about a revision of values, a revision of the constituent elements of the house."[1]

Le Corbusier's grand utopian visions have been corrupted through economic rationalism and urban design empty of imagination and responsibility and devoid of "values" (by Corbusier's definition or anyone else's). In *The Language of Post-Modern Architecture*, Charles Jencks sums up the failure of modern architecture: "Modern architecture died in St Louis, Missouri, on July 15, 1972, at 3:32pm (or thereabouts) when the infamous Pruitt-Igoe scheme...were given the final coup de grace by dynamite. Previously it had been vandalised, mutilated and defaced by its... inhabitants, and although millions of dollars were pumped back, trying to keep it alive...it was finally put out of its misery. Boom, boom, boom."[2]

The Pruitt-Igoe scheme was a common mutation of Corbusier's Freehold Maisonettes, which now dot the landscape throughout cities worldwide. But don't blame Corbusier. In fact, he was right; he just went about it the wrong way. Where the modernist utopian vision ultimately fails is where Corb V2.0 picks up the pieces. Now, as housing prices soar and the word *sustainability* threatens to be overrun by the word *survivability*, designers and builders need to utilize the ubiquitous technology still flooding our cities to create dense housing that people actually want to live in. Instead of turning the house into Corbusier's machine, Maynard uses the machine to erode social hierarchy and flatten real estate economics.

Corb V2.0 takes well-designed apartments (rather than badly scaled containers) and uses modern infrastructure to deal with the areas where apartment blocks fail (i.e., social hierarchy and lack of adaptability or responsiveness). Through the mobility afforded by shipping equipment, the utopian ideal is once more employed in a housing solution, which Corbusier dreamt of back in '23.

Within Corb V2.0 the spatial hierarchies traditionally determined by wealth and the implied status wealth evokes are dissolved, real estate values become flattened, and a new lifestyle alternative (already adopted in mobile technologies such as phones and laptops) begins to emerge in housing.

The mobility that Corb V2.0 allows also gives the residents an unprecedented degree of control over their social environment; this reengineered shipping container, preprogrammed for real living, establishes a feedback loop of user's responses to density, orientation, and height. This is fuzzy logic on a grand scale.

1 Le Corbusier, *Towards a New Architecture*, trans. Frederick Etchells (London: John Rodker, 1931; Mineola, NY: Dover, 1986), 227.

2 Charles Jencks, *The Language of Post-Modern Architecture*, 4th rev. ed. (1977; repr., New York: Rizzoli, 1984), 9.

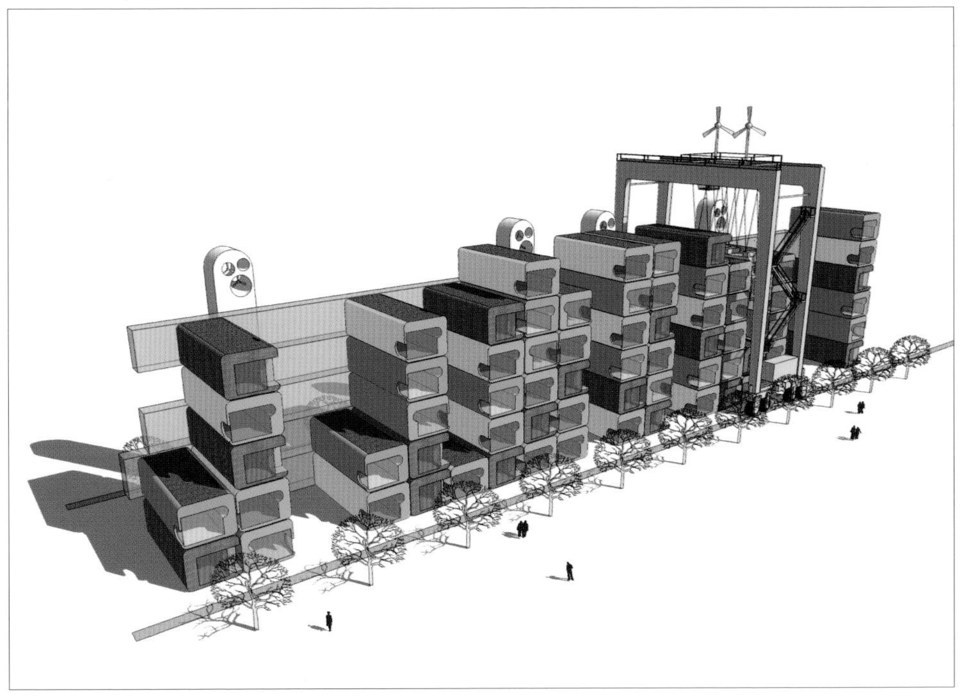

WOULDN'T IT BE GREAT IF WE WERE APARTMENTS RATHER THAN BORING CONTAINERS?

THAT'S A RAD IDEA.

WHY DO ARCHITECTS KEEP TRYING TO SQUASH HOUSES INTO CONTAINERS? CONTAINER DIMENSIONS ARE TERRIBLE. WHY NOT DESIGN A KICKARSE APARTMENT AND USE ALL OF THE OTHER FUN TOYS THAT WE FIND ON DOCKS TO HELP DEAL WITH THE MANY TROUBLING ISSUES THAT THE MODERNIST VISIONS OF DENSE HOUSING HAVE DIFFICULTY ADDRESSING?

HUMAN ARE NOMADIC IN NATURE. ON AVERAGE AUSTRALIANS WILL LIVE IN 14 DIFFERENT HOMES DURING THEIR LIFETIME. THIS NOMADIC TENDENCY IT USUALLY DESIRED, NOT IMPOSED. OUR NOMAD LIFESTYLE IS TYPICALLY BY CHOICE, CITING BOREDOM, STATUS ENVY, CONTINUOUS CHANGE IN FAMILY DYNAMIC, GROWTH OF PERSONAL WEALTH, CHANGE OF WORKPLACE ETC..

POSTMODERN HOUSING (IN MORE WAYS THAN ONE)

THIS WEEK, I AM THE PENTHOUSE, YIPPEE!

I'M GOING TO BE ON THE GROUND FLOOR FOR A FEW DAYS.

THAT'S OK THOUGH, ITS ONLY FOR A FEW DAYS.

CORB V2.0 OFFERS THE MODERIST UTOPIA, WITH A LOT LESS BAGGAGE :
-LESS HEIRACHY
-NO FIXED VIEW
-NO FIXED ORIENTATION
-A CHANGING URBAN FORM
-A SEASONALLY ADJUSTABLE HOUSING DEVELOPMENT
-ADJUSTABLE SOCIAL DYNAMIC
-INTERCHANGEABLE NEIGHBOURS
-FLAT PROPERTY AFFORDABILITY

IF YOU DECIDE TO MOVE THEN YOU CAN SIMPLY THROW YOUR HOUSE ON A TRUCK OR SHIP AND HAVE IT WAITING FOR YOU WHEN YOU ARRIVE.

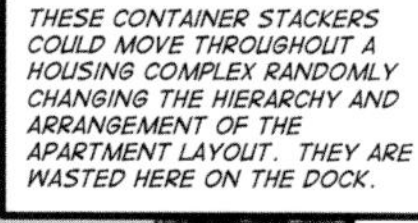

SSA TERMINALS

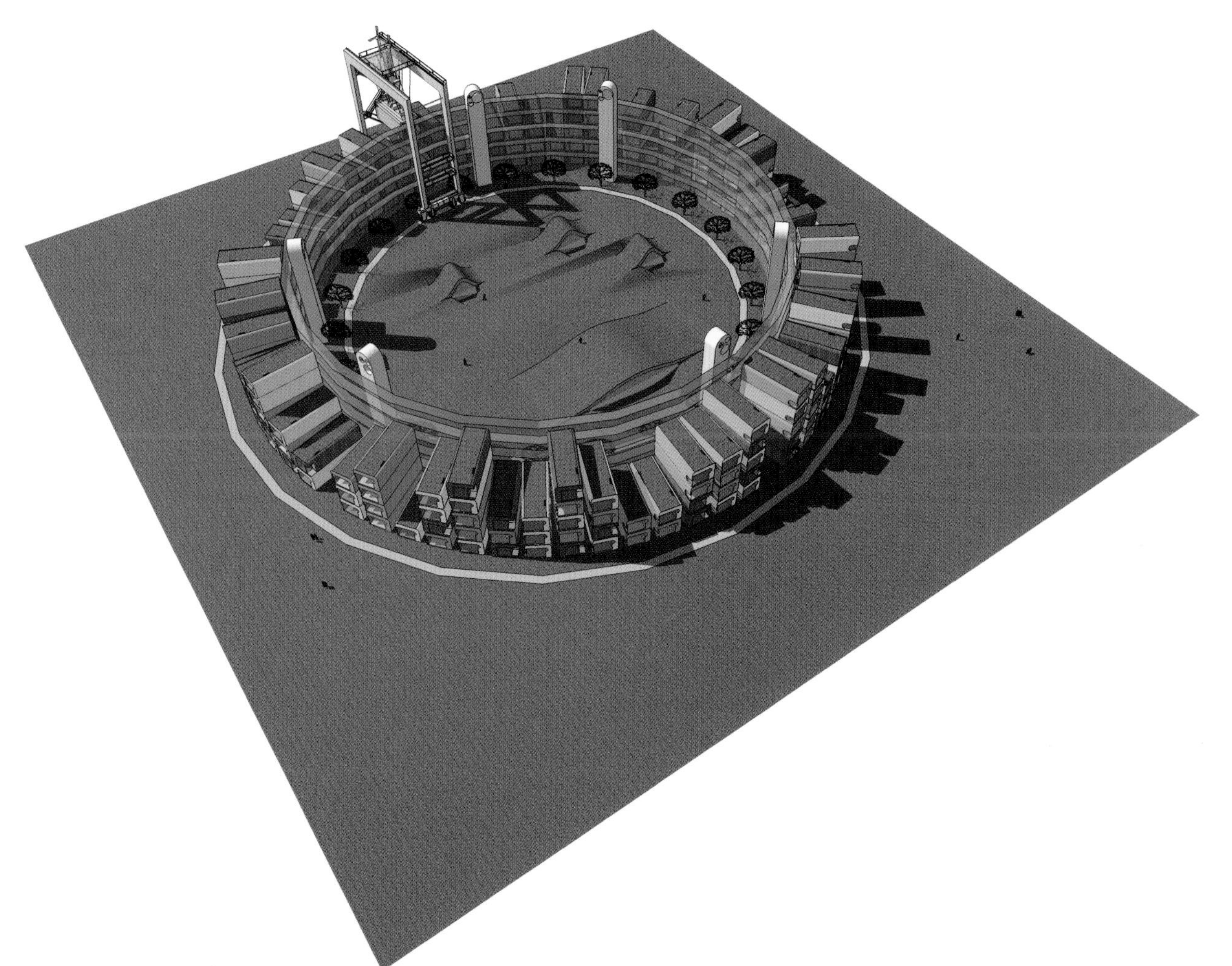

Design Pod

It is a reality of the changing work/studio environment that we are working in larger, shared, nonhierarchical spaces for the sake of efficiency. Many feel the loss of their autonomy and power over their own actions. By using technology to create a sense of ownership and territory, the individual is empowered, while a free and fluid work/studio environment is maintained.

The Design Pod offers the individual the opportunity to control the work environment and to manipulate and move the work station into any environment desired. In the case of the architect, engineer, landscape designer, or any professional who carefully considers site, environment, and local conditions, the Design Pod provides the perfect opportunity for on-site design and documentation and also correspondence to home office, consultants, or the client.

The Design Pod has a super-lightweight carbon-fiber frame with internal rubber seat padding and an external rubber shell. The structure allows the user to transport the Design Pod easily, while its rubber shell protects it from damage and allows the user to personalize the space through the choice of pattern, color, and texture. Providing the full spectrum of multimedia output within the Design Pod, the Design Desktop gives the individual control over the everyday tools of the design studio, such as communication and basic computing needs. More importantly, though, the Design Desktop provides the user with the ability to bridge the gap between virtual drawing and design technologies and the designer's mind. It does this by removing the mouse from the equation.

The Design Co-Op

Much like the Design Pod in its use, the Design Co-Op is an open studio for a much larger office environment. The Design Co-Op was established by Maynard's office staff for use by similar workers, such as artists and designers, to pool their resources and produce an environment of shared space, media, and ideas.

The Design Co-Op is owned by all and, like the contemporary work environment, it is nonhierarchical. The individual stores the Design Pod here and has access to all the Design Co-Op's resources twenty-four hours a day.

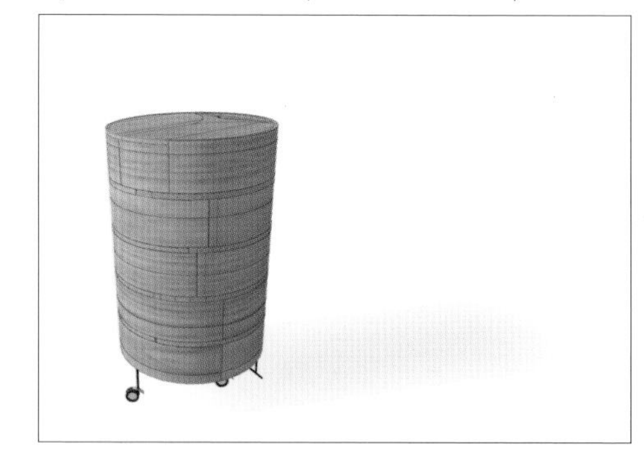

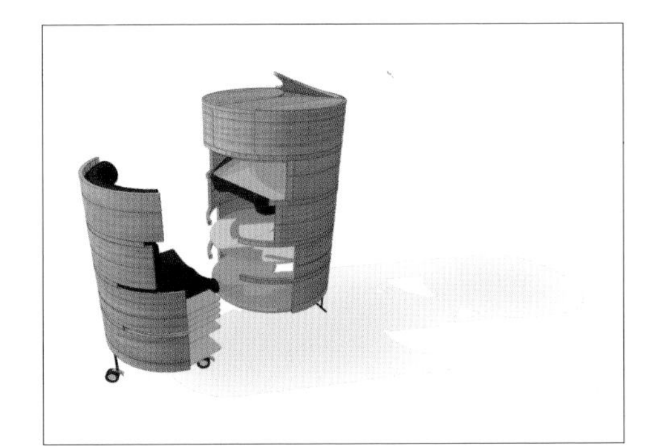

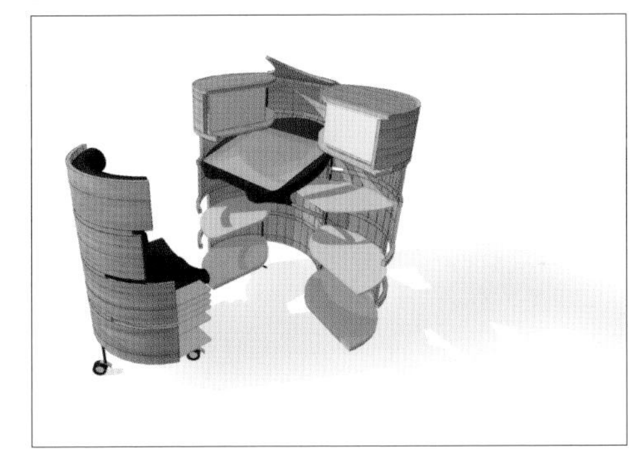

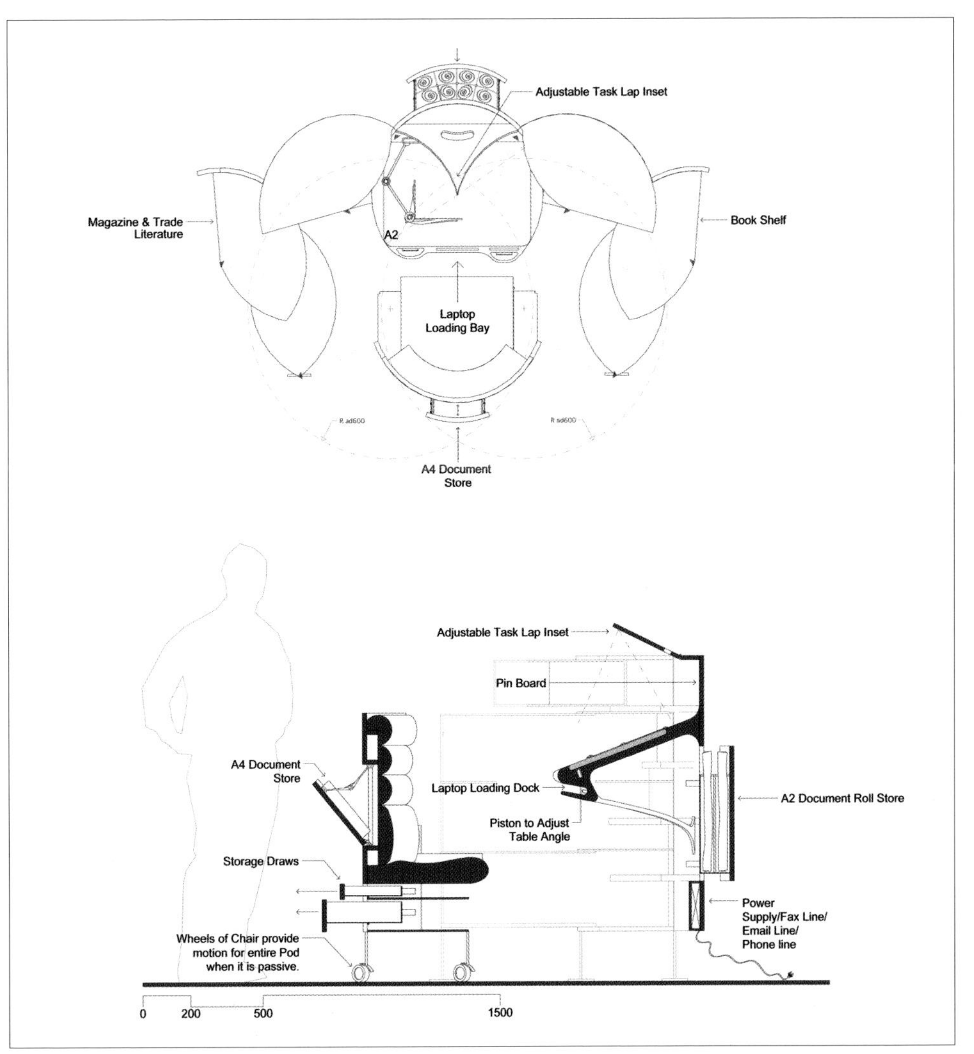
Adjustable Task Lap Inset
Magazine & Trade Literature
Book Shelf
A2
Laptop Loading Bay
A4 Document Store
Adjustable Task Lap Inset
Pin Board
A4 Document Store
Laptop Loading Dock
A2 Document Roll Store
Piston to Adjust Table Angle
Storage Draws
Power Supply/Fax Line/ Email Line/ Phone line
Wheels of Chair provide motion for entire Pod when it is passive.
0
200
500
1500

Andreas Vogler

Andreas Vogler, a Swiss architect, was born in Basel in 1964. Vogler studied art history and literature and worked as an interior architect before he attended architecture school at the Swiss Federal Institute of Technology, graduating in 1994. From 1995 to 1996 he worked with Richard Horden and Christoph Ingenhoven. From 1996 to 2002 Vogler was an assistant professor at the University of Technology in Munich. He was a leading force, combining teaching with research to create fully realized buildings. In 1998 Vogler initiated and led a design semester for aerospace architecture, during which students planned for the International Space Station and a human Mars mission, for which they performed parabolic test flights at NASA's Johnson Space Center.

During his time at the University, Vogler submitted several prize-winning competitions entries. In 2003 he founded the company Architecture and Vision with his Italian partner Arturo Vittori. The office concentrates on innovative concepts in architecture and technology transfer from aerospace; its investigations include a study for a mobile embassy for Switzerland, as well as for inflatable tents for the European Space Agency. The firm built a prototype of the tent named DesertSeal in collaboration with ESA and the Italian company Aero Sekur. The prototype was first exhibited at the Museum of Modern Art in New York City and recently became part of the museum's prestigious permanent collection. Vogler was visiting professor at the Royal Academy of Fine Arts School of Architecture in Copenhagen, researching prefabricated houses from 2003 to 2004, and is currently collaborating with the Delft University of Technology in the Concept House research program. He lectures and teaches internationally, including commitments in Switzerland, Denmark, Hong Kong, and Italy, has been organizing scientific conferences, and is currently working on a study for a space simulator. He is a member of the Bavarian Chamber of Architects, the Deutscher Werkbund, and the American Institute of Aeronautics and Astronautics. Says Vogler, "By leaving our planet in the past forty years we learned more about it than in the four thousand years before."

Andreas Vogler

DesertSeal

DesertSeal is an inflatable tent for extreme environments that makes use of the temperature curve in hot, arid regions, where the air gets considerably cooler the more distant it is from the Earth's surface. This effect is used by many desert animals, not least by the camel. An electric fan, powered by a flexible solar panel and batteries, constantly blows cooler air from the top of the tent into the body of the livable space within. The tent consists of an air-beam structure made of yellow polyurethane-coated polyethylene fiber; and its awning is a silver-coated high-strength textile that reflects heat and protects from direct sunshine. The beauty of this configuration derives from its functionality and efficiency, particularly in dealing with such natural energies as sun and wind. A newly developed solar film will be tested for additional energy gain to power the electric fans. The aero-space-architecture background of the designers is visible in the conception, construction, and materialization of the project, which is derived from natural resources and has minimal weight for transportability.

The DesertSeal project was sponsored by the European Space Agency and involved collaboration with numerous international groups, including the European Astronaut Corps, in Cologne, Germany; Aero Sekur s.p.A., in Aprilia, Italy; and VHF Technologies SA in Yverdon-les-Bains, Switzerland.

DesertSeal 600

MarsCruiserOne

MarsCruiserOne (MCO) is a design prototype that would accommodate a human exploration of the red planet; it is based on a less developed concept by EADS Astrium Bremen for the European Space Agency. It consists of a pressurized compartment for habitation, exploration, and scientific investigation and a mobility system based on extra-large, adaptable wheels. The concept is a modification of the European Space Agency's Ariane launch vehicles (or AR++), which are being developed for future planetary exploration. The pressurized area is designed to provide all functions of a habitat for two to three astronauts for up to twenty days until they drive back to the base station, where they can resupply their food, air, and energy. MCO's interior is designed and optimized for the living and working environment of the crew, while most of the vehicle's subsystems, such as its power supply and environmental conditioning system are located outside the main structural shell. Studies of mobile homes, boats, and aircraft interiors informed the design and helped identify potential technology gaps. An optimal and multifunctional use of space has to be achieved. The cockpit is designed for driving as well as for sleeping; the seats provide optimal ergonomic support for both functions; and the space has to perform the transition from operational working space to private space.

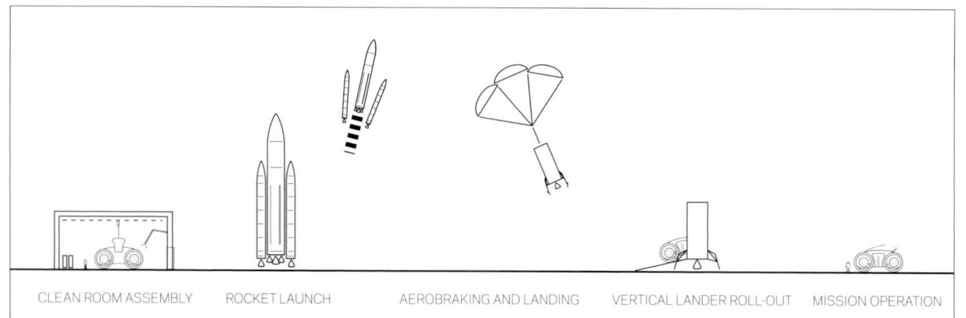

The MCO could be the starting point of a permanent human settlement on the red planet, allowing the first small scientific outpost to grow into a more and more articulate complex, where infrastructure and habitats will give shape to the first new city in the solar system.

MoonBaseTwo

MoonBaseTwo is an inflatable laboratory for the Moon that will allow the crew to conduct scientific research onsite and explore the extreme environment. Designed to be transported in the new ARES V rocket fairing, it is self-inflating and ready to be inhabited in conjunction with the landing of the first astronauts.

MoonBaseTwo is destined to serve as the main habitat integrated into the currently discussed NASA "Constellation" program for a human mission to the Moon. The inflatable base will be packed into a 24′ 6″ diameter and 32′ 8″-long module. After inflation, the external bags will be filled with regolith, or lunar dust, to provide radiation protection for the astronauts. Three additional modules complete the main habitat with airlocks and a rover docking port. MoonBaseTwo will draw its energy from a self-contained solar plant. Currently, the rim of the Shackleton Crater on the Moon's south pole is discussed as a possible site for the lab. Polar sites receive most light on the Moon and thus avoid the fourteen-day-long nights experienced at other regions on the Moon (a full Moon day equals thirty Earth days).

In such an environment, psychological and social issues become an important design driver. A clear division between private and common space is employed by arranging soft materials and large volumes inside; this allows for higher vertical movement in low-gravity situations and creates a sense of space in this extreme environment. The base floor contains the public spaces, and private crew quarters hang from the ceiling of the dome. The soft, smooth, and friendly shapes improve and enrich the daily visual and haptic perception of space in such a highly technical environment. LED lighting technology, which can be changed in color and intensity, will be used to re-create terrestrial sequences of night and day, enhance the perception of space, and counteract the sensory deprivation of the astronauts.

MoonBaseOne
ThalesAlenia

MercuryHouseTwo

The MercuryHouseTwo prototype continues the modernists' quest for better and brighter living in the future. The factory-built five-story house is a critical alternative to the excessive land use of the traditional single-family home, using new building technologies, smart materials, and efficient, space-saving designer sense.

The materials revolution fostered by nanotechnology will result in an evolution of the architecture and paradigm of the house. Within twenty years the control of material properties by applied nanotechnology and embedded systems will revolutionize conventional building skins. Already today vanadium oxide can be applied in films so thin that glass stays transparent but can actively control the passage of infrared rays and, as a result, can fine-tune temperature controls. Solar cells will be transparent, and a robot-maintained inflatable greenhouse will reduce trips to the grocery store.

The organization of the house is vertical—like a tree. Comprising prefabricated modules delivered from a factory, the stories are stacked using a muscle-driven lift. Furniture is integrated in an industrial-designed aircraft style, and big buoyant wheels allow for fast removal in case of emergency situations. After site installation, water is filled into the hull to provide weight and wind resistance. Recently updated concepts for the MercuryHouseTwo design include monitoring systems for small children and possibly a system that evolves into a ply structure for older kids.

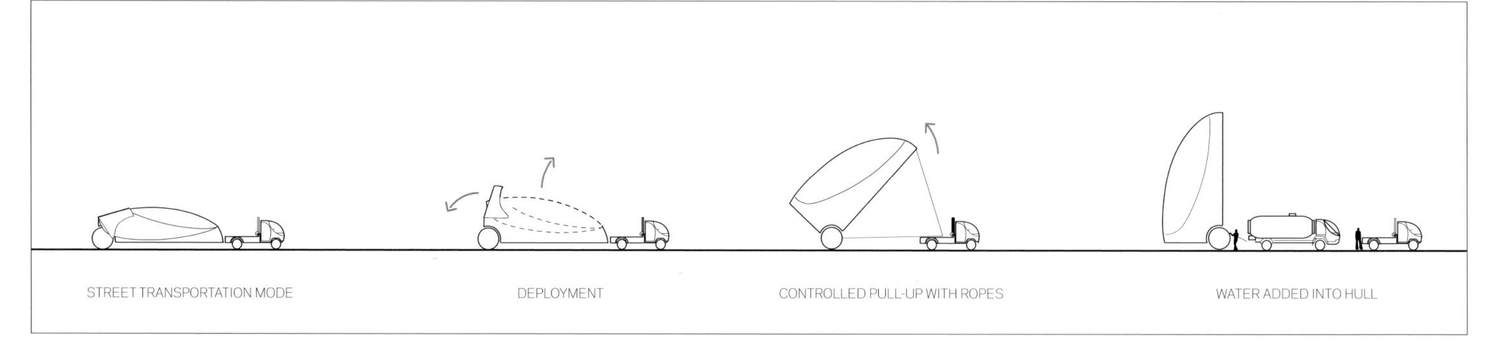

INDUSTRIES

Horden Cherry Lee Architects

Horden Cherry Lee Architects is an architectural, planning, and design studio led by Richard Horden, Stephen Cherry, and Billie Lee. Founded in 1999, the practice quickl y earned a reputation for designing and delivering high-quality architecture in the commercial and residential sectors, as well as developing expertise in master planning and product and furniture design. Projects include the award-winning micro-compact home (2005) and Ercol Furniture Factory (2002) and high-tech city office schemes for Borchard Lines (2005), Royal Bank of Scotland (2005), and the refurbishment and reconfiguration of Shield House (2002), which secured listed building consent for the property. The collaborative approach and conceptual development of the micro-compact home, which was successfully launched in Munich in 2005, reflects the practice's commitment to developing highly efficient alternatives to existing architectural typologies based on technological advances.

Horden Cherry Lee Architects strives toward the pursuit and interpretation of lightness, producing buildings that "touch the earth lightly" by virtue of an aesthetic of delicate, highly glazed frameworks, which open up interior spaces to the landscape and minimize intrusion on nature. The practice's design approach is entirely holistic, with this aesthetic extending to bespoke furniture and fittings, including a range developed with engineers from British Aerospace's Concorde, Rolls-Royce factories, and the all-aluminum "System 26" furniture system.

Founding partner Richard Horden is also University Professor at the Faculty for Architecture at the Technical University of Munich and heads the Institute for Architecture and Product Design. Billie Lee has been a visiting teacher at the Antwerp School of Architecture and is a visiting professor at the Technical University of Vienna, where he has run annual design workshops since 1999.

micro
www.microcompacthome.com
Vschinauncha
da Silvaplana

The Carbon Neutral Micro-Compact Home

The Carbon Neutral Micro-Compact Home (M-CH) is a carbon-neutral home designed for a Swiss client in the Alps that uses a myriad of off-the-grid energy producing technologies to power the entire structure. At an exceedingly high elevation, facing south with a clear horizon, Micro-Compact Home Ltd. has worked with Ove Arup and Partners and Ernst+Basler Ag Engineers in Zurich to develop the project in detail. This version of the M-CH runs all-electric (as opposed to natural gas or coal for heating, cooling, and cooking) and is powered by photovoltaic solar panels with a vertical wind generator. Daytime excess power is diverted into the larger community grid, and nighttime power is provided by the wind turbine and reserve batteries. Heating and air-conditioning is ducted to each of the four function spaces—eating, sleeping, washing, and working. Long-duration LEDs light the interior and external walkways. The carbon-free home also contains a few green-energy perks like the lockable ski and snowboard drawer accessed from the outside and included in the insulation zone, so skis and boots are warm in the morning.

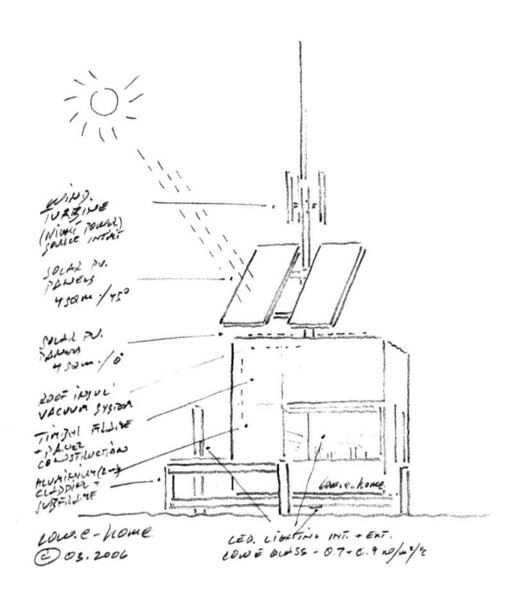

O2

Micro-Compact Village

This village features seven micro-compact dwellings, each transportable and lightweight, combining high technology and low energy use, much like the Smart Car.

Specially selected students of the Technical University of Munich were chosen to live in six of the cubes for the 2005–6 winter semester alongside their British professor, architect Richard Horden, who masterminded the design of the Micro-Compact Home.

The concept was developed over four years by Richard Horden and Horden Cherry Lee Architects with the German practice Lydia Haack + John Hoepfner Architects. Lydia Haack teaches alongside Horden in the university's Institute of Architecture and Product Design, and their students worked on the M-CH project, as did students at the Tokyo Institute of Technology, where Professor Horden also taught.

The layout is influenced by traditional Japanese teahouse architecture. Internally the space is divided into zones. A compact zone of wet services houses the toilet, shower, and the kitchen. On the central axis are the entrance and kitchen-circulation area, which also serve as access to seating in the lower dining area. The upper-level sleeping bunk (for two) can be folded out of the way, while below, the sunken dining area can also double as a second sleeping space (also for two).

Lightweight technology is used throughout, including insulated vacuum aluminum paneling mounted on a timber-and-galvanized-aluminum section that provides the basic structure. Several units can be mounted on an external aluminum frame in vertical and horizontal formations, around central lift and stair cores to form a "village."

The Micro-Compact Home is ideal for business travelers, holiday homes, or other short-term residential or academic uses. It requires no furniture and comes with all integrated energy and communications systems. Raised off the ground, it has a minimal impact on its environment.

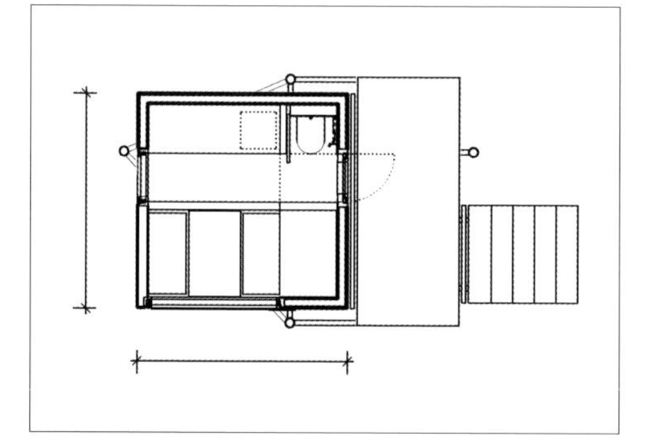

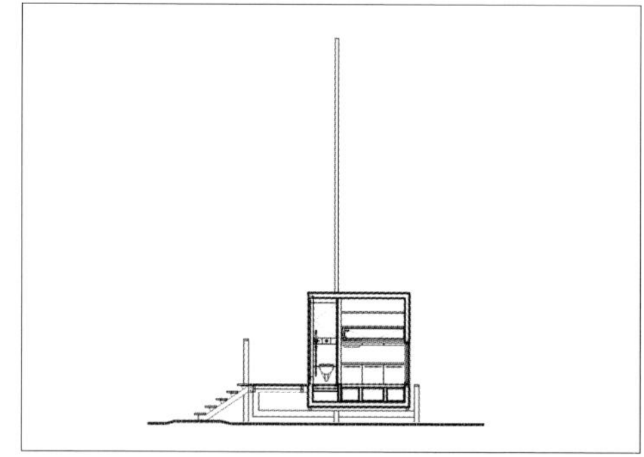

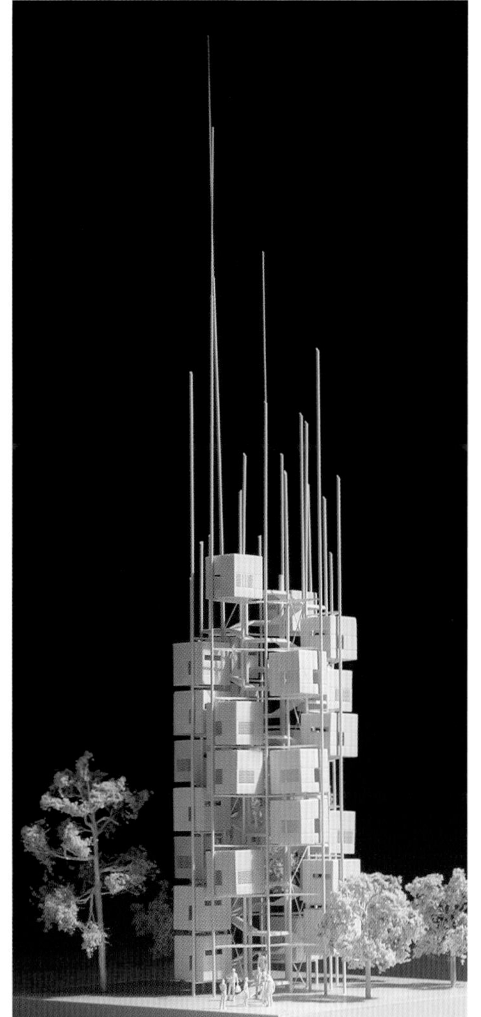

can jump

can touch
can laugh
can join
can jump
can hide
O2
O2

N55

In 1994 a noncommercial exhibition space and lab was initiated in Nørre Farimagsgade 55, Copenhagen. N55 grew out of this collaboration. In 1996 a number of persons started living together in an apartment located in the center of Copenhagen, trying to "rebuild the city from within" and using their everyday life as a platform for public events and collaborations. In the year 2000, Floating Platform and N55 Spaceframe were constructed in the harbor area. N55 Spaceframe served as a starting point for local initiatives and interventions, a work and living space for the group until 2004; it now serves only as a living space.

N55 is a platform for persons who want to work together, share places to live, economy, and means of production. N55 is based both in Copenhagen and in LAND. This design collective publishes its own manuals for the reproduction of its designs in the spirit of open-source software sharing. As a result, N55 designs are implemented in various situations around the world, whether initiated by N55 or in collaboration with different persons and institutions.

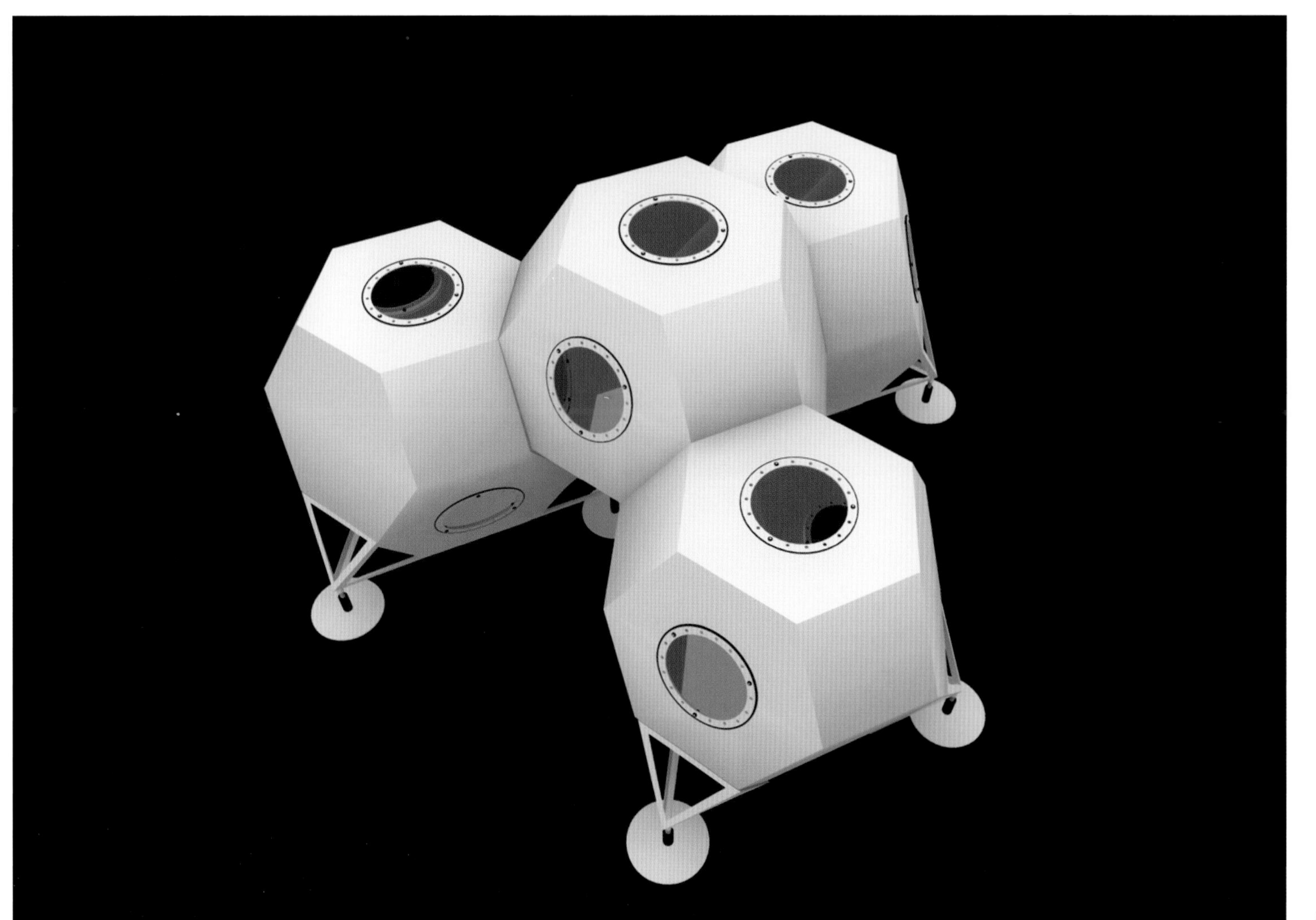

Floating Platform

Floating Platform is a modular construction that functions as a buoyant foundation for N55 Spaceframe or for other lightweight constructions. This base system can be rearranged and can be used for other purposes, such as building up landmass.

Floating Platform is a space lattice comprising small modules made from stainless steel and built-in buoyant tanks that can all be assembled by hand. The modular system facilitates gradual extensions and makes the platform less vulnerable to damage; for example, leaks only have local impacts and can be serviced locally, thus alleviating whole-system repairs.

The platform is constructed as an "octet truss" space lattice and is shaped as an equilateral triangle. One hundred eighty-nine polyethylene tanks, concentrated in the three pontoons situated in each corner of the platform, make the platform buoyant. The pontoons are constructed from three layers of tanks and are shaped as tetrahedra, with one vertex pointing downward and a plane facing upward.

The floating system can also be shaped according to the intended usage. Floating Platform has a low net buoyancy, which gives it an economic advantage, but limits the amount of weight that can be added to the platform. However, the modularity of the platform allows for the possibility of adding more buoyancy when there is a need for it: extra tanks may be connected to the sides of the platform, material may be put into the cavities in the steel truss, or extensions can be made from the same building system. Tanks for water and wastewater, as well as toilets, can be integrated in ways that do not add load to the construction.

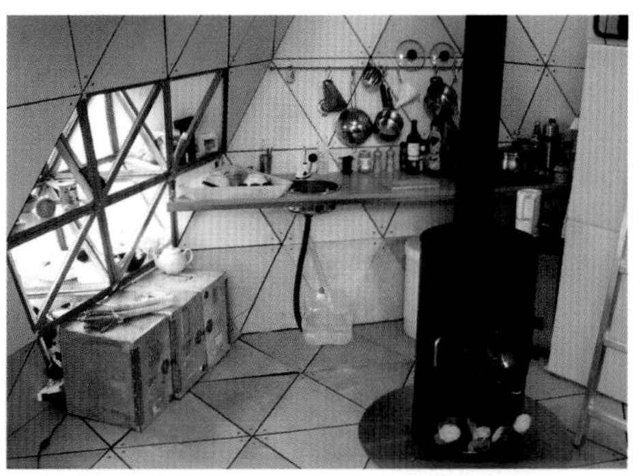

Micro Dwellings

Micro Dwellings is a system for making low-cost dwellings of variable sizes for any number of persons consisting of movable housing modules that can form different configurations on land, on water, and under water. The system allows for a diversity of materials, as well as changes and adaptations to the design over time.

Micro Dwellings are modular, which allows them to be stacked up, rearranged, or gathered together with other systems into small communities. The Micro Dwellings can be built onto rooftops of existing buildings or suspended from a bridge or a wall. The modules can be mounted on wheels for mobilizing or connected to form floating constructions. Micro Dwellings can also be made into watertight, amphibian houses that can be completely submerged or partly elevated near or at the water's surface.

Most functions are built into the dwellings' walls, and furniture, along with household equipment, is provided by moveable elements that change functions during the day. Supply modules can be mounted on the outside of the main modules.

Micro Dwellings are able to adapt to changes in life, e.g., residents moving in and out, the arrival of children, and an expanding need for space. Conversely, certain modules of the dwelling can be taken away if living spaces need to contract. People who want to live together can simply let their dwellings grow together. Likewise, it is easy to separate modules and move them if that is desired. Micro Dwellings do not in themselves define a social constellation, but they provide the basic equipment people need to configure their own social setting. The present version of the system is made of cheap steel plates and can be constructed by anybody who knows how to weld.

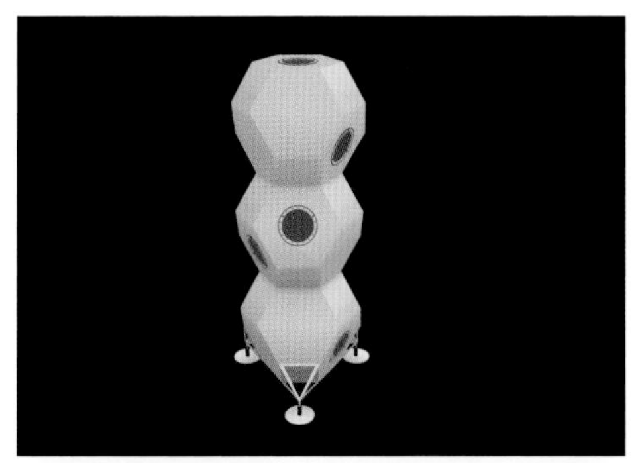

Small Truck

Small Truck is a lightweight, low-cost, man-powered vehicle that enables people to move loads up to about 650 pounds at slow speeds. Transporting things using a Small Truck, the driver gets physical exercise while working; combining these activities saves time for other things. Small Truck provides shelter from wind and rain and can be equipped with various systems: platforms and trailers that can support micro-economical initiatives, like a transport company, a small shop, restaurant, cinema, or office. It could also be equipped to offer an unfoldable concert stage with scenery, public library, or a mobile home.

N55 has placed directions for construction of the Small Truck on its website and hopes that individuals will build their own. N55 likes the idea of many different designs of the vehicle, so that the Small Truck might be available to others.

The chassis of Small Truck was realized through a collaboration between Pelle Brage and N55. This means Small Trucks will appear in various materials and shapes, depending on the context in which they are meant to function.

Snail Shell System

The Snail Shell System is a low-cost system that enables mobility for people in various environments, since it is operable both on land and water. One unit supplies space for one person, and one person can move the unit slowly, either by pushing it like a wheel or walking inside it or on top of it.

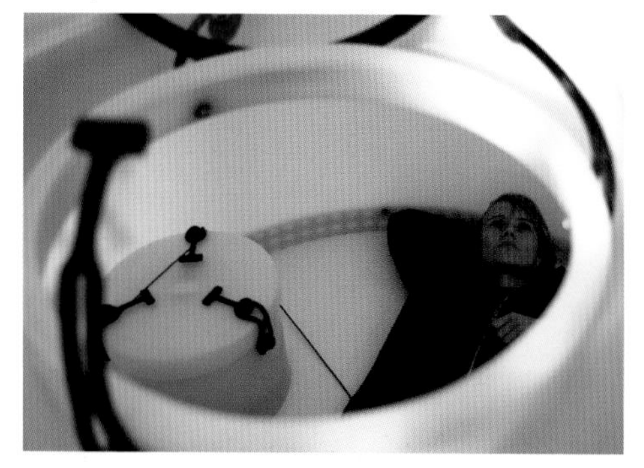

On water it can be rowed, towed by a kite, or hooked up to a vessel, such as a ferry. The unit rests on one flat side and can be anchored in lakes, rivers, harbors, or at sea. On land it can be placed in city spaces, fields, forests, or parks. The Snail Shell System takes up very little space and can easily be placed into a setting in a discreet way. It can be buried in the ground, exposing only the entrance. It can also function as a comfortable space inside existing buildings.

Several units can be joined to form temporary communities where the unit can be hooked up to existing infrastructure like telecommunication lines and electricity cables by connecting it, for example, to street lamps. If special solar, wind-turbine, or thermal insulation devices are added, the unit can supply its own energy.

The Snail Shell System can also be used as a means of storage or as a moving unit for shipping small loads. In light of its subversive and collective tendencies, N55 would also like to point out that the Snail Shell System can provide protection for persons when they participate in demonstrations or protests.

Atelier Bow-Wow

Atelier Bow-Wow was established by Yoshiharu Tsukamoto and Momoyo Kaijima in 1992. The pair's interest lies in diverse fields, ranging from urban research to architectural design and the creation of public artworks.

Its research has been compiled into books concerning the thirteenth arrondissement of Paris and work in Japan, focusing on, among others, Tokyo, Atami, Mito, Fukuoka, and Tsukuba.

The practice has designed more than twenty private houses in central Tokyo, and in recent years the completion of projects such as the Hana Midori Cultural Center and Mado Building mark an expansion toward larger-scale works such as museums and commercial architecture.

Atelier Bow-Wow's experimental projects with micro public space have been exhibited across the world in international art museums as part of events such as the Gwangju Biennale (Korea), Shanghai Biennale (China), Echigo Tsumari Triennale (Japan), Venice Biennale (Italy), Busan Biennale (Korea), São Paulo Biennale (Brazil), and Yokohama Triennale (Japan).

Atelier Bow-Wow's seminal report on small architecture in urban Tokyo, *Pet Architecture Guidebook*, and its analysis of Tokyo's anonymous hybrid architecture, *Made in Tokyo*, were both published in 2001, marking the firm's initial moves into urban research. In 2006 the practice's research concerning Tokyo and the changes since the economic-bubble collapse was published in *Bow-Wow from Post Bubble City*, and its comprehensive monograph, *Graphic Anatomy: Atelier Bow-Wow*, was published in 2006. Atelier Bow-Wow's first solo exhibition How to Use the City at Kirin Plaza, Osaka, was followed in March 2007 by Atelier Bow-Wow: Practice of Lively Space—Global Detached Houses and Micro Public Space, held at Gallery Ma, Tokyo.

White Limousine Yatai

A *yatai* (street cart) serves food at the street side and has a charming ability to tie people together and encourage interaction. The standard yatai is about five feet long and designed to be run by a single person. Inspired by this moveable meeting place, Atelier Bow-Wow began to design its own "limousine yatai," stretching the prototype to thirty feet, enabling more people to gather around it while elevating its urban appeal.

The long yatai, painted white, visited various events and facilitated a micro public space at each stop. Every time it turns a street corner it creates a small traffic jam, but its humorous appearance makes people laugh. To be a true yatai, the limousine had to vend a food product, and an assortment of all-white or translucent food—Japanese rice wine, tofu, pickled white radish—was served.

Furnicycle

This project was designed for the Shanghai Biennale in 2002, which was themed "Urban Creation." Atelier Bow-Wow's interpretation of the theme sought to mutate ways in which urban development influences art and architecture. Surprised by the density of high-rise buildings in Shanghai, Atelier Bow-Wow realized that what was most dynamic about the city were the living spaces people made just anywhere by putting furniture on the street, and also the customized bicycles for carrying large packages and goods that move all around the city.

With Furnicycle Atelier Bow-Wow connected furniture and bicycle back-to-back in order to use the optimal performance of each object. By changing the way individuals might deal with these city fluxes, the goal became to create a new phenomenon on the street. Chaircycle, Bedcycle, and Tablecycle can operate separately or join to make a living space on the street. When taken out and tested at a street-corner tea stand in Shanghai, the Furnicycle attracted more people than expected, and a micro public space emerged.

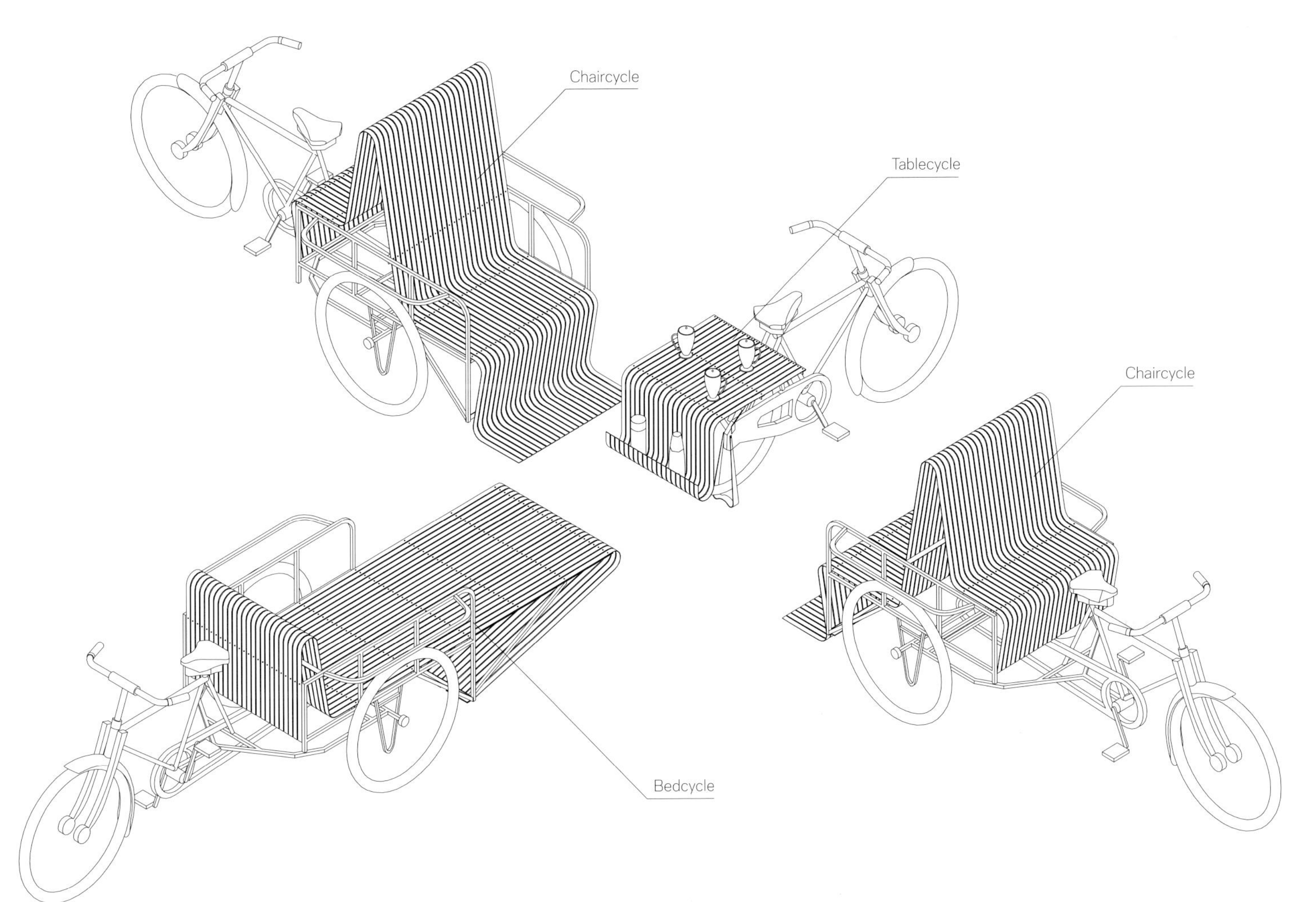
Chaircycle
Tablecycle
Chaircycle
Bedcycle

School Wheel

The bed of the Oncheoncheon Stream in Busan City, South Korea, has been paved over in concrete, creating a relaxation and recreation space for the residents of the surrounding area. Water still runs through a central channel in the concrete riverbed, and when the river swells, the water level rises and once again swallows the open space. Depending on the amount of water in the river, the area appears and disappears, creating a public space that is governed by the characteristics of variability and even redundancy. This rising and falling action was exploited for the purpose of an exhibition as part of the Busan Biennale in 2006, and an installation was created that allows one to enjoy the experience of such an ephemeral public space. One big blackboard, forty-eight chairs, an awning, and lights make up the School Wheel, a temporary classroom situated under the blue sky. The whole front surface of this portable structure was painted as a blackboard and could be used as a satellite classroom for any nearby educational institutions. Here anyone could become a teacher and anyone could become a pupil.

Following the installation period in Busan, the School Wheel moved to the Museum of Twenty-first Century Art, in Kanazawa, Japan, to become part of Atelier Bow-Wow's exhibition Lively Projects in Kanazawa, for which the architects invited people with knowledge or experience of Kanazawa to give lectures in the School Wheel.

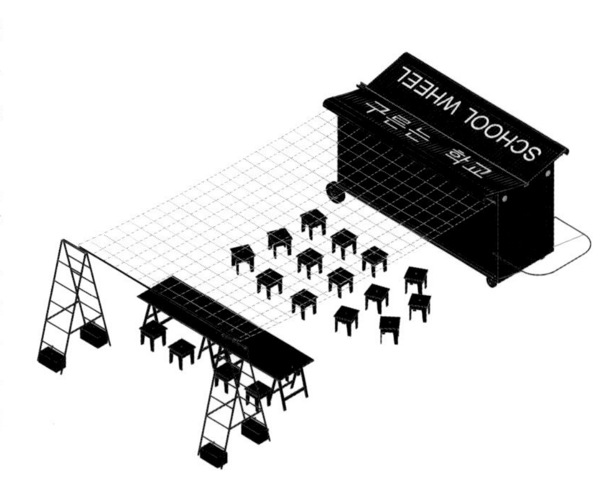

SCHOOL WHEEL
구르는 학교

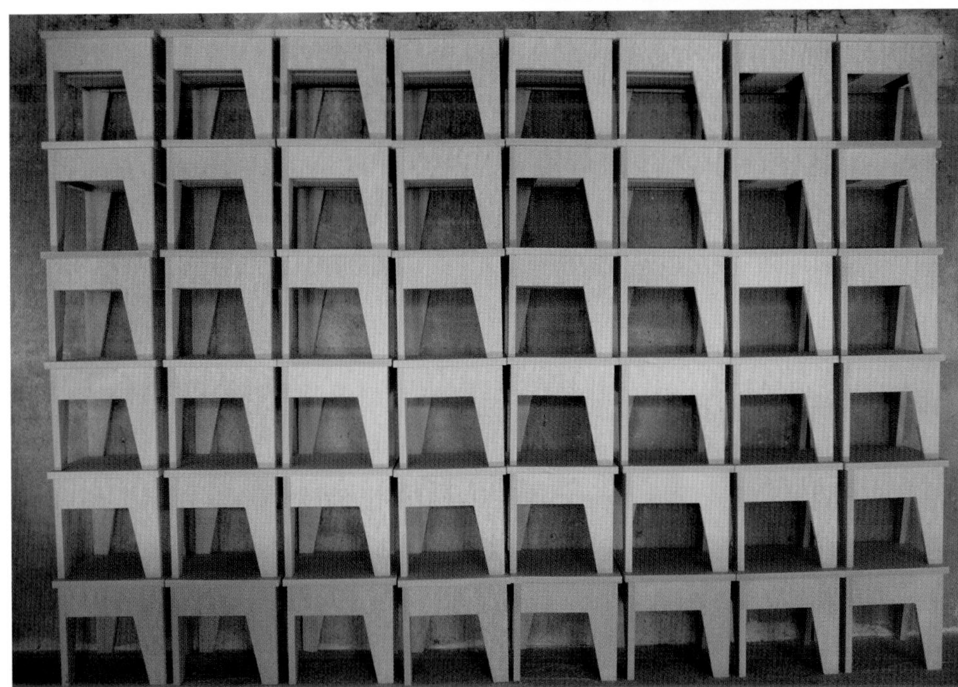

The Mark Fisher Studio

The Mark Fisher Studio has an international reputation for spectacular live-entertainment design. Fisher has created some of the most memorable rock concerts ever staged, including *The Wall* for Pink Floyd, *Steel Wheels* and *Bridges to Babylon* for the Rolling Stones, and *Zoo TV* and *Popmart* for U2. He also designed the theater and scenery for *KÁ*, the permanent show by Cirque du Soleil at the MGM Grand in Las Vegas, and the scenery for *We Will Rock You*, the Queen musical that has been running worldwide since 2002.

Working with Peter Gabriel, Fisher designed and co-directed *OVO*, an acrobatic show at the Millennium Dome in London that ran for 999 performances through the year 2000, reaching an audience of 6.5 million people. He was also creative director for the opening celebrations at the dome, an event that featured a cast of more than a thousand performers, including four hundred carnival artists.

The Mark Fisher Studio has designed concerts for many of the most famous names in popular entertainment. Past and current clients include the Japanese artists Yumi Matsutoya and Mr. Children; the European artists Johnny Hallyday, Mylène Farmer, Herbert Grönemeyer, and Marius Westernhagen; and many British and American artists, including Elton John, Tina Turner, Robbie Williams, Janet Jackson, Phil Collins, and Cher. The studio has also designed live-music TV shows for such broadcasters as the BBC, ITV, MTV, and VH1.

In addition, the studio has designed numerous one-off events and fixed installations, among them four NFL Super Bowl halftime shows; spectacles worldwide for Jean Michel Jarre; *IllumiNations 25* for Walt Disney World in Orlando; *Aquamatrix*, the nightly show at the Lisbon Expo '98; and the Turin Winter Olympics Opening and Closing Ceremonies in 2006.

Genesis *Turn It On Again* World Tour 2007

The tight logistical and economic constraints of touring-rock-concert stages mean that the utilitarian aspects of their construction frequently dominate the form of the finished structure. For their 2007 tour, Genesis wanted their stage to have, in place of expediency, an anti-industrial, organic quality that could reflect the understated Englishness of their music. In addition, the European portion of the tour was booked in blocks of three and four nights back-to-back. While this is normal (if rather hard work) for an arena show, it was unprecedented for a bespoke stadium tour traveling with a custom-built stage. The final design was an efficient structure that concealed the rectilinear primary structure beneath expressive curvilinear surfaces.

To meet the logistical challenge of such a compressed time frame, the design was based around four complete systems of conventional stage components that leapfrogged between venues. The production equipment—lighting instruments, LED screens, public-address systems, and band gear—typically arrived at the stadium on the day of the show, ten hours before the audience was admitted. To make the transition flow quickly, almost all work was undertaken

at stage level. The design ensured that the various production departments could occupy the stage without competing for space. This meant that each department could lift its equipment into the air without waiting on other departments to finish and move out of the way. The lighting equipment on the curved finger trusses was loaded at stage level and pulled up dedicated tracks. The 250-foot-wide curved LED screen—a custom fabrication of Barco OLite tiles arranged in varying densities across pressed-aluminum battens—was designed to be erected in under three hours. It hung from a curved fascia truss that itself contained an array of preprogrammed lighting scenes. The truss, which contained all the rigging for the screen, was guided upward from below stage level on pre-built towers. Once it was in position, the carts of LED panels were positioned underneath and the screen lifted into place.

HELSINGIN PANTTI

Robbie Williams *Close Encounters* European Tour 2006

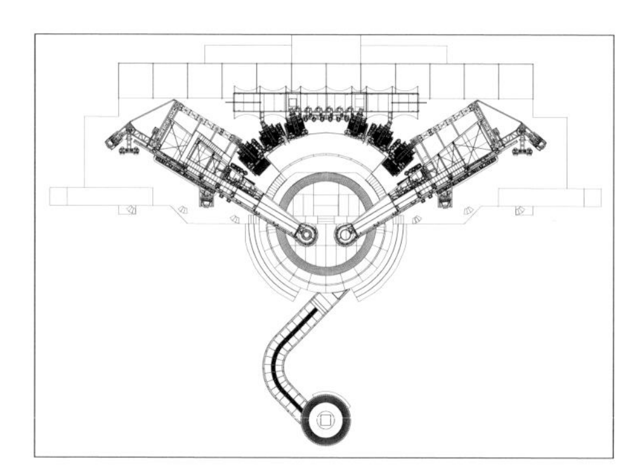

This stage design created a heroic background for Robbie Williams's *Superhero* performance, which climaxed with his making an encore entrance by gondola from one of the cantilevered arms. The thrusting masculine forms framed a rich multimedia environment of video screens and precise lighting configurations. A small main stage and even smaller side stage sat within the frame.

The show traveled with two sets of primary steelwork, consisting of black steel towers and truss supports. To minimize the work at its finished destination onstage, the yellow custom-built cantilever structures were assembled at ground level and craned into place. They reached out fifty feet over the main stage. One set of rented lighting, public-address, and video equipment was installed during the final twenty-four hours before each show. The lighting fixtures in the bellies of the cantilever trusses were attached to secondary trusses at ground level and lifted into place using chain hoists. To maintain a clean look for the audience, all back-line technicians worked from weatherproof bunkers beneath the stage.

The Rolling Stones *Bigger Bang* World Tour 2005–7

Early conversations about the design for the Rolling Stones *Bigger Bang* tour focused on ideas for placing the band in an "operatic" environment. The first sketches explored decadent, picturesque fantasies of grand nineteenth-century opera-house decor. These studies were distilled to incorporate the audience into the staging behind the band in "boxes." As the design developed, the boxes metamorphosed into sweeping expressionist balconies that flanked a large high-resolution LED video screen, forming a streamlined back wall to the stage. Upstage of the balconies, panels of low-resolution LED video created a luminous backdrop to silhouette the opera audience.

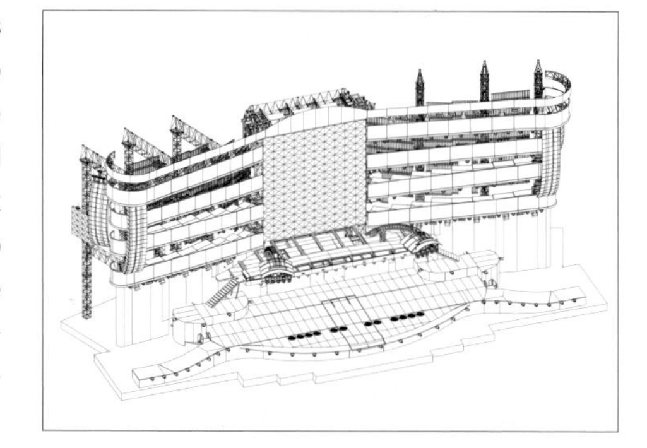

The core structure was organized into two trajectories of masts and trusses that supported a series of cantilever brackets to provide bearing for the audience balconies and lifting points for the fascia panels. The primary steelwork was mostly composed of stock stage components, stabilized by a custom base frame of I beams set out beneath the three-foot-high substage. Custom-fabricated cantilever brackets and audience decks were integrated into the primary steelwork, which took approximately thirty hours to erect. Along with these three sets of primary steelwork, one set of fascia panels, video screens, and other show components traveled from venue to venue for fit-up during the final twenty-four hours before the show.

AMERIQUEST
MORTGAGE COMPANY
AMERIQUEST
MORTGAGE COMPANY

SPORTS AUTHORITY
UNDER ARMOUR
FENWAY

MMW

MMW Architects was founded by Magne Magler Wiggen in the fall of 1997. The first project to put the office on the map, nationally and internationally, was Philtex—a mobile home made from two steel containers.

Since then MMW has made its mark with new and innovative solutions that challenge the conventions of building. The Time Machine, Caravantex, Villa Bakke, Kilden Cinema, Bølgen & Moi restaurant, Inferno, and Saboteur are examples of this.

The office consists of nine people who possess a broad competence in the fields of architecture, planning, building conservation, stage design, and furniture design. And although MMW's first and foremost endeavor is that of an architectural practice, it also has a gallery, studio 34, which regularly hosts art and design exhibitions.

In 2001 *Wallpaper** magazine listed MMW as one of the ten most exciting architecture and design practices in the world. The practice has also been recognized with lectures at various institutions, such as the Architectural Association, London; the Oslo School of Architecture; the Norwegian University of Science and Technology, in Trondheim; and the Bergen Association of Architects and Norsk Form, in Oslo. The practice was the only Norwegian participant in the Image02 exhibition in London in the spring of 2002, and it has had several of its projects presented on television and radio, as well as in publications like *Frame*, *AMC*, *Quaderns*, *Dwell*, *Design*, *Fjords*, *Architectural Review*, and *Byggekunst*.

Working with temporary buildings and semipermanent structures has inspired the firm to produce realistic solutions for spectacular architecture within limited time frames and budgets, proving that fantastic architecture doesn't have to come at a high cost and can be achieved by challenging established construction methods. MMW has well-earned experience in bridging the gap between the conceptual and the real. The office also has a strong understanding of established construction methods and was awarded the prize for excellence in use of concrete by the Norwegian Concrete Association in 2003.

Magne Magler Wiggen is professor at AHO, the Oslo School of Architecture and Design, and is focusing on developing new methods for temporary and lightweight uses for concrete. He is also in charge of diploma studios at AHO, committed to establishing innovative methods for architecture and design.

GAD

Commissioned by its owner, Alexandra Dyvi, this temporary gallery made from steel and glass was designed in response to the shipping containers that have traditionally occupied this kind of area—a burgeoning port town. The shipping-container shape is one of not only utility, but also of history. The gallery's site is the home of some of Oslo's oldest shipyards, and it contains a timeline of Norwegian seacraft from nineteenth-century fishing vessels to today's hulking industrial shipping cranes.

The doors to the new gallery, called GAD, were to open less than one year from the initial meeting with the client, but the tight schedule wasn't the only design problem. The biggest design challenge came from the owner's requirement that the gallery move periodically with only a few weeks' notice. A self-contained infrastructure is what finally allowed for safe and efficient movement of the unit.

To give GAD an open feeling and to let in plenty of natural light for the artwork within, MMW placed circular windows opposite each other across the length of each rectangular room, so as to bathe the gallery in northern light. Conventional finishing techniques and materials (insulation, sheets of plywood, and Sheetrock layers) were used within the modules to achieve the look needed to display art—that of the classic white cube.

GAD

The Frog and Harmonica

As part of the Norwegian National Museum of Art, Architecture, and Design's opening in 2005, MMW was commissioned to develop a holistic architectural solution to generate more public interest around the events.

The temporary pavilions celebrated the hundredth anniversary of the nation-state of Norway and highlighted the city of Tullinløkka as the future arena for the new National Museum.

The Frog, the temporary main pavilion, was the largest structure designed by MMW. The structure's concept was based on the pneumatic principle by which a self-supporting construction is erected by establishing higher air pressure inside the pavilion membrane, as with a beach ball or an air bed.

The outer membrane walls, made from an opaque PVC weave with a fire-retardant layer, glow with a green color. The inside walls are covered with a white surface, allowing for projections of film and images.

Adjacent to the Frog, the Harmonica, a smaller, semipermanent pavilion, worked as a vital exhibition hall and as a design graphic telling about the future museum building that would soon be situated there.

fremtiden

framtiden
the

LOT-EK

LOT-EK is a design studio based in New York City, founded in 1993 by Ada Tolla and Giuseppe Lignano. Since then LOT-EK has undertaken residential, commercial, and institutional projects both in the United States and abroad, as well as exhibition design and site-specific installations for major cultural institutions, including the Museum of Modern Art, the Whitney Museum of American Art, the Guggenheim, and the New Museum of Contemporary Art.

Tolla and Lignano earned master's degrees in architecture and urban design from the Universita' degli Studi di Napoli, Italy, and completed postgraduate courses at Columbia University, in New York. Besides heading their professional practice, they are currently teaching at the Columbia University Graduate School of Architecture. They also lecture throughout the United States in major universities and cultural institutions, including Princeton University, Yale University, Rice University, Rhode Island School of Design, and the Guggenheim Museum, as well as overseas at ETH, Zurich; ETSAB, Barcelona; IUAV, Venice; Bartlett and Royal College of Art, London; Wits University, Johannesburg; and Tokyo Design Block.

The office has achieved high visibility in the architecture, design, and art worlds for its innovative approach to construction, materials, and space; for the use of technology as an integral part of architecture; for addressing issues of mobility and transformability in architecture; and for blurring the boundaries between art, architecture, and entertainment. Its projects are published in national and international publications, among them: the *New York Times*, the *London Times*, the *International Herald Tribune*, the *Wall Street Journal*, *Wallpaper**, *Domus*, *A+U*, *Interior Design*, *Wired*, *Surface*, *Metropolis*, *Vogue*, and *Graphis*. LOT-EK's first monograph, *Urban Scan*, was published by Princeton Architectural Press in February 2002. *Mixer*, published by Edizioni Press, came out in 2000, and *Mobile Dwelling Unit*, published by DAP, came out in June 2003.

UNIQLO Container Stores

UNIQLO, Japan's most popular apparel retailer, commissioned LOT-EK to design and construct these "container stores," taking the form of a ship's cargo container directly from the pier to customers in the streets, bringing the best of UNIQLO to all New Yorkers in a unique and efficient way. The twenty-foot-wide-by-eight-foot-deep container stores weigh approximately fifteen thousand pounds fully loaded. This includes merchandise, IT equipment, and the container itself.

LOT-EK orchestrated the loading, lifting, and shipping of the stores, which arrived at specific Manhattan locales on a flatbed truck; the modules were then lifted via crane and placed onto the street.

ユニ
クロ
UNI
QLO

EXIT>
UNI
QLO

Mobile Retail Unit DIM

Much like UNIQLO's container stores, the Mobile Retail Unit DIM consists of a fifty-three-foot-long truck that triples in width by expanding at both sides with the push of a button. When the unit is traveling, the side expansion pods are pushed in and the floor folds up, indicating the location of central stationary areas where the parts of the permanent store are situated. All the clothing stacks, accessory drawers, fitting rooms, and lounge seats pull out of the central container. Display of the clothes is concentrated at the thin layer of flat-screen monitors, creating an infrastructure for the virtual video display of the rapidly updated merchandise that is, or will soon be, available.

Monitors are organized in vertical columns attached to the front of pullout drawers, which contain the clothes displayed on them. As they pull out the drawers to inspect the merchandise, customers can take a snapshot of their own face via a mounted camera, the images are collected throughout the day and replayed randomly on the upper monitors.

The fitting rooms drop from the ceiling to free valuable space when not in use, and the cash register is a portable handheld machine.

To enhance the constantly animated display systems inside, reflective surfaces cover the interior walls. On the exterior, reflective surfaces make this project contextual, leaving the inside-outside connection to be filtered through the brand's logo, which wraps around the object.

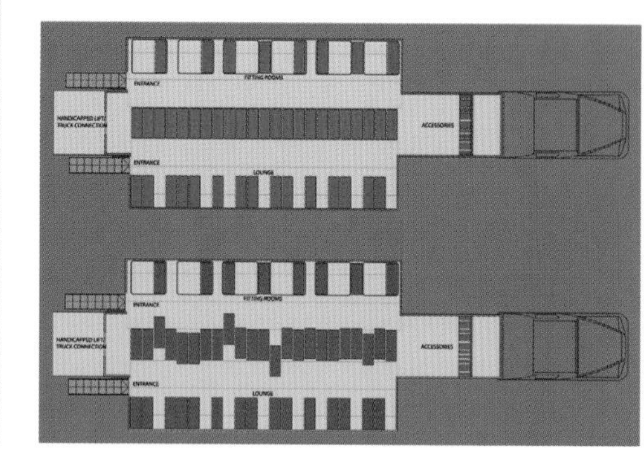
FITTING ROOMS
ENTRANCE
HANDICAPPED LIFT/
TRUCK CONNECTION
ACCESSORIES
ENTRANCE
LOUNGE
FITTING ROOMS
ENTRANCE
HANDICAPPED LIFT/
TRUCK CONNECTION
ACCESSORIES
ENTRANCE
LOUNGE

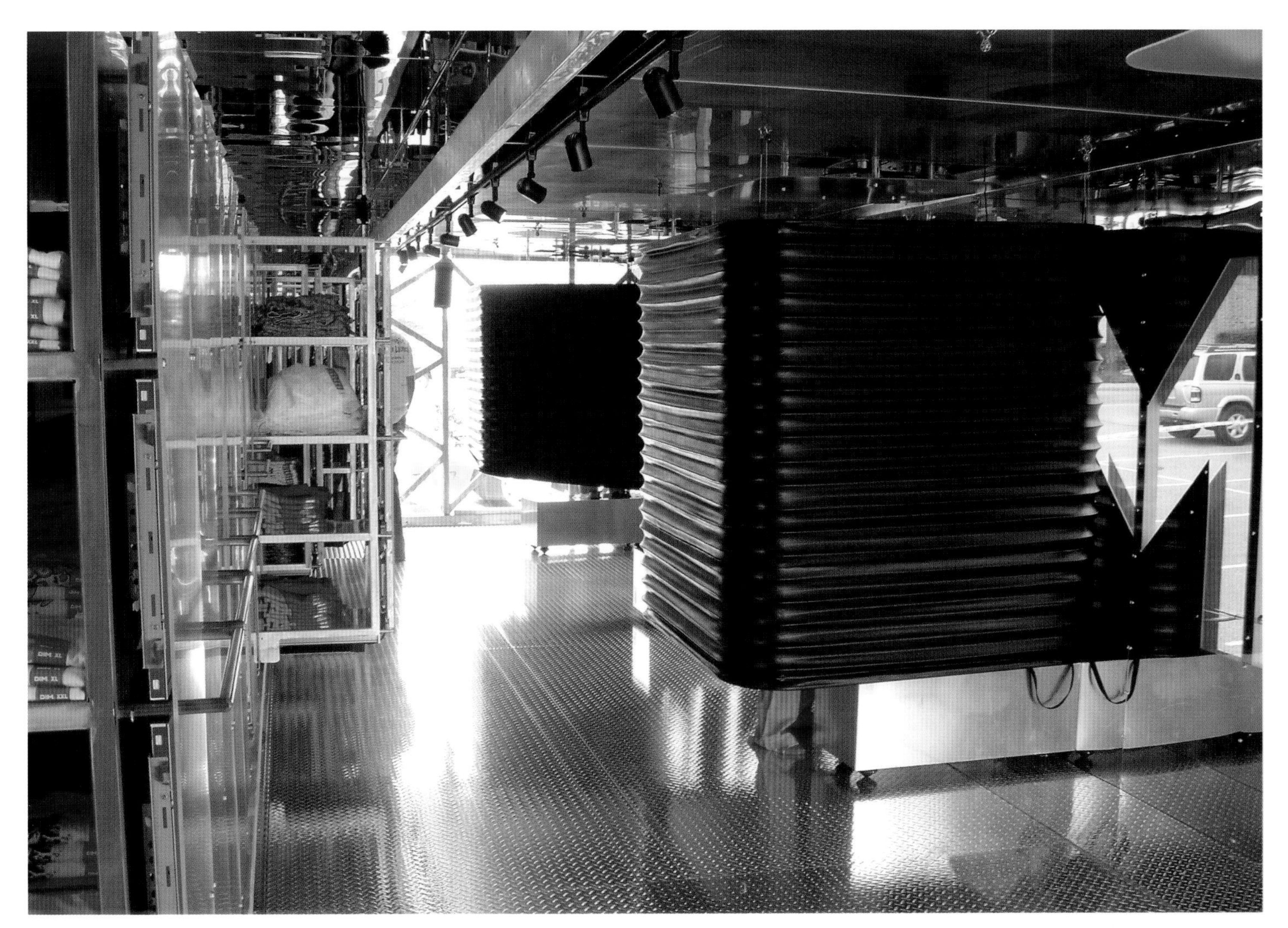

Office of Mobile Design

Jennifer Siegal is founder and principal of the Los Angeles–based firm Office of Mobile Design (OMD), which is dedicated to the design and construction of modern, sustainable, and precision-built structures. She is known for her work in creating the mobile home of the twentieth century.

She earned a master's degree from the Southern California Institute of Architecture (SCI-Arc) in 1994 and was a 2003 Loeb Fellow at Harvard's Graduate School of Design, where she explored the use of intelligent, kinetic, and lightweight materials. In 1997 she was the architect-in-residence at the Chinati Foundation and in 2004 a fellow at the MacDowell Colony in her hometown, Peterborough, New Hampshire. Presently she is the inaugural Julius Shulman Institute Fellow at Woodbury University, she is also the editor of *Mobile: The Art of Portable Architecture* (2002) and the founder and former series editor of *Materials Monthly* (2005–6), both published by Princeton Architectural Press. A monograph on Jennifer Siegal was published in 2005.

Her innovative design sensibilities and expertise in futuristic concepts, prefabricated construction, and green building technologies were recognized by the popular media in 2003 when *Esquire* named her one of the design world's "Best and Brightest" and the Architectural League of New York included her in the acclaimed Emerging Voices program. She was featured in *Fast Company*'s "Masters of Design" in 2006 for her exceptional approach to utilizing new material and forms to create architecture. She was honored when mayor Antonio Viaragosa presented her with the History Channel's 2006 Infiniti Design Excellence Award for her entry in the Los Angeles City of the Future 2106 competition. Her most recent built project, the Country School, which is the first green prefab school in Los Angeles, was recognized as one of the five best buildings in Los Angeles in 2007.

Abbot Kinney
OVERSIZE LOAD
UP21601

Seatrain Residence

This three-thousand-square-foot custom residence playfully uses traditional commercial and industrial materials. Using storage containers and steel found on-site in downtown LA, Office of Mobile Design created an oasis without abandoning or disguising the industrial landscape that inspired the design and provided the materials.

Situated by the Brewery, a live-work artists' community, the house's large panels of glass open up the space, allowing natural light to pour in and visually connecting it to the rest of the community. In keeping with the spirit of the community in which the house is being built, the project was a collaborative experiment between the client, Richard Carlson, and the fabricators. The design/build approach allowed for creative and structural decisions to be made as the house was being constructed.

This home grew up from the land around it, engaging with and incorporating the industrial history of downtown LA through the use of raw detritus. Grain trailers became a koi pond and a lap pool. The large storage containers create and separate the dwelling spaces within the house. Each storage container has its own individual function: one is the entertainment and library area; another is a dining room and office space, overlooking the garden below; another serves as the bathroom and laundry room; and yet another is the master bedroom, a visually dramatic protruding volume that wraps around the upper part of the house. This unfussy space allows for the dynamic interplay of materials and forms; the contrast of corrugated metals, industrial containers, and exposed wooden beams are all highlighted with warm, calm green hues.

All of the containers used in the house have been altered in surprising ways. Some have been severed into separate pieces, while others have been added onto, layered, or wrapped, showing the myriad design possibilities available when repurposing these materials. There are wrapped design elements throughout the house, including a twelve-foot-high steel-plate fence around the entire site. At one point it lifts up, stretching to become a canopy that gives shade to the entrance, creating the feeling of the ground plane being tilted upward. Here recycled materials are not just practical and cost effective, but they also create a unique, dramatic architectural vocabulary. The innovative combination of recycled storage containers, grain trailers, steel, and glass result in a house that is highly sculptural, open, and LA modern.

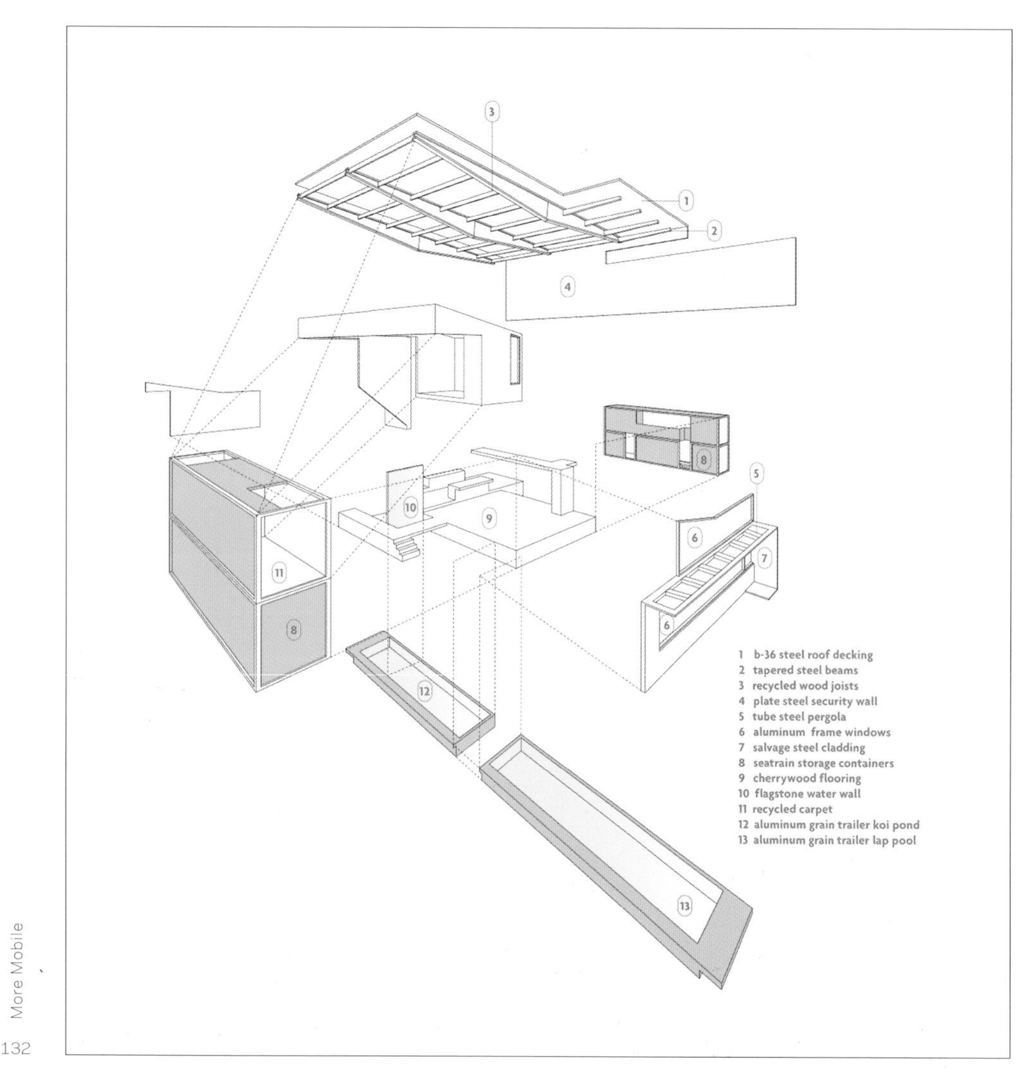
1 b-36 steel roof decking
2 tapered steel beams
3 recycled wood joists
4 plate steel security wall
5 tube steel pergola
6 aluminum frame windows
7 salvage steel cladding
8 seatrain storage containers
9 cherrywood flooring
10 flagstone water wall
11 recycled carpet
12 aluminum grain trailer koi pond
13 aluminum grain trailer lap pool

Globetrotter

Reviving the tradition of traveling Shakespearean troupes of the seventeenth century, OMD—working in conjunction with Shakespeare Festival/LA—has designed a mobile, modular vehicle that is able to transform into a fully equipped theater on any relatively flat site.

Measuring twelve feet wide by fifty feet long by fourteen feet high, the vehicle would be hitched to a typical truck cab. Once parked in its temporary location, it would be deployed to present a stage that is roughly the same size as the stage at the original Globe Theatre. Wing walls unfold from the sides to enliven acoustics, support lighting, and provide scenery surfaces and projection surfaces for filmed close-ups. Backstage six pneumatic PneuPods inflate to provide dressing rooms, a production office, and a ticket/concession space, plus sleeping areas for the cast and crew. Along the interior spine of the vehicle are sinks, showers, toilets, and equipment racks.

Photovoltaic panels on the roof provide power for sound and lighting. One side of the vehicle contains three large LED screens, which advertise the traveling troupe with scenes from previous performances as the truck rolls down the road.

This crossbreeding of high theater and high camping would generate a carnival-like atmosphere and entertain a wide variety of changing audiences.

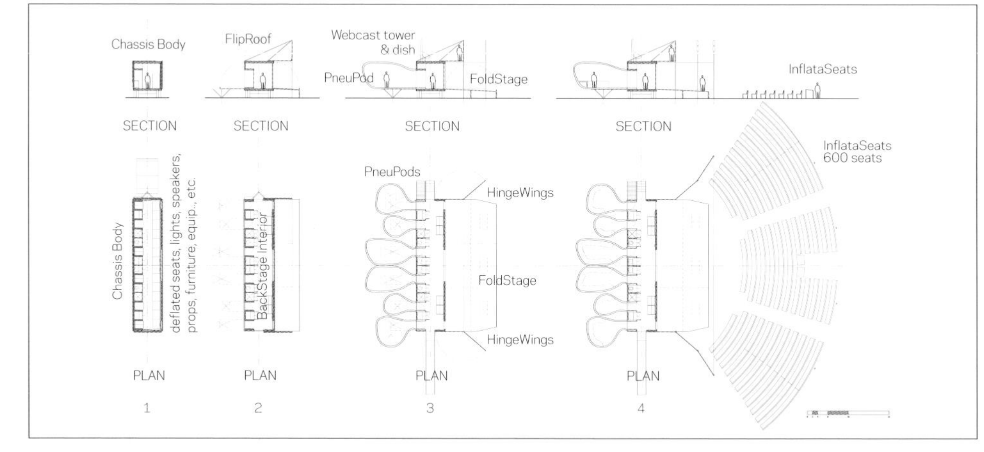

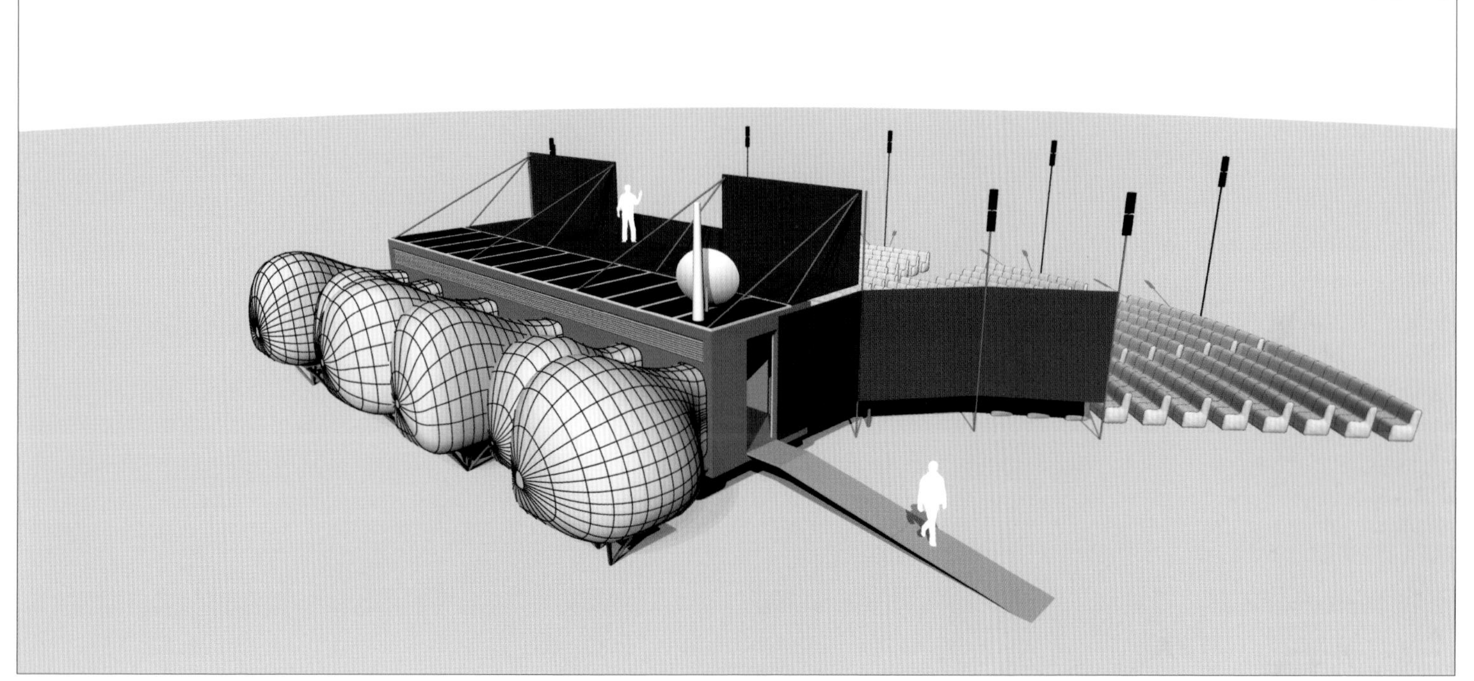

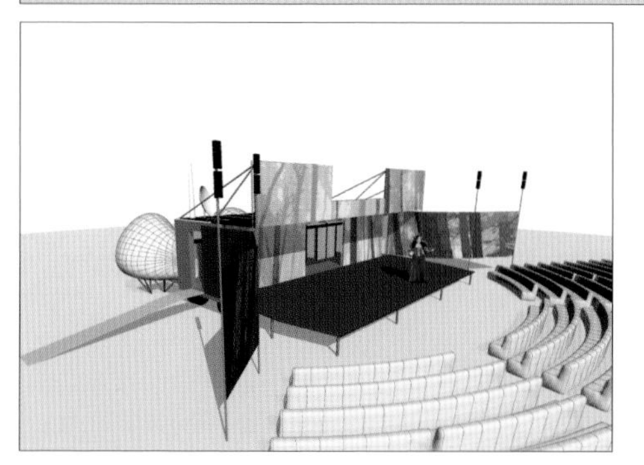

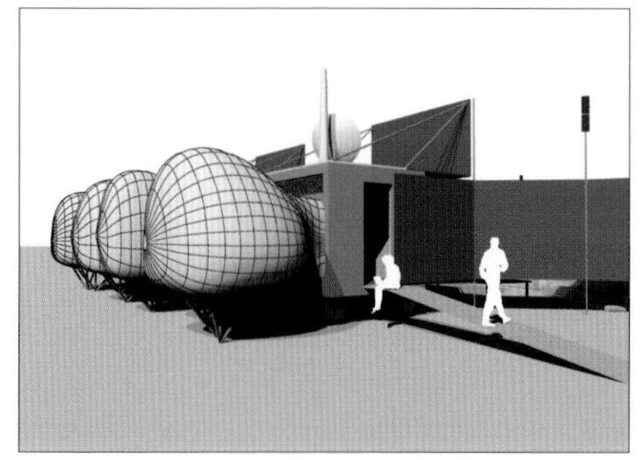

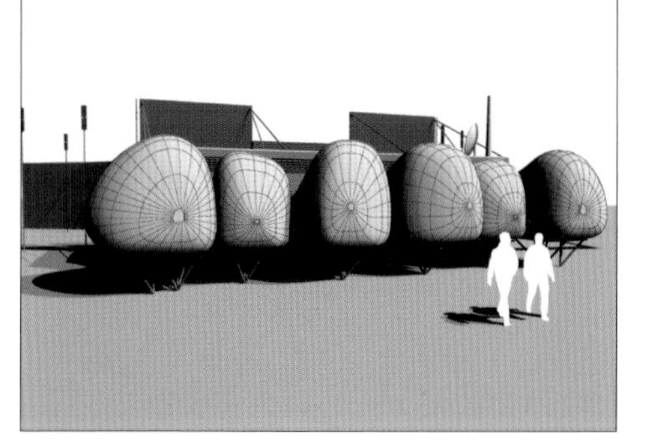

Hydra 21

The buoyant survival structure Hydra 21 provides a temporary ocean refuge for citizens of war-torn nations. Distributed by helicopter, the prefabricated units inflate on impact with the ocean's surface. The structure's outer skin, a synthetic rubber reinforced with a glass-lattice fabric, lets in ample light and protects inhabitants from temperature extremes ranging from -20°F to +120°F. Pliant polymers flex when photovoltaic energy is applied via pumped seawater to a desalinization system; treated wastewater returns to the ocean. Solar cells and a system that harnesses wave energy supply Hydra 21 with electricity. A floating exterior garden, anchored to the marine dwelling's outer skin, gives modern-day ocean dwellers a place to grow food and to stretch their legs.

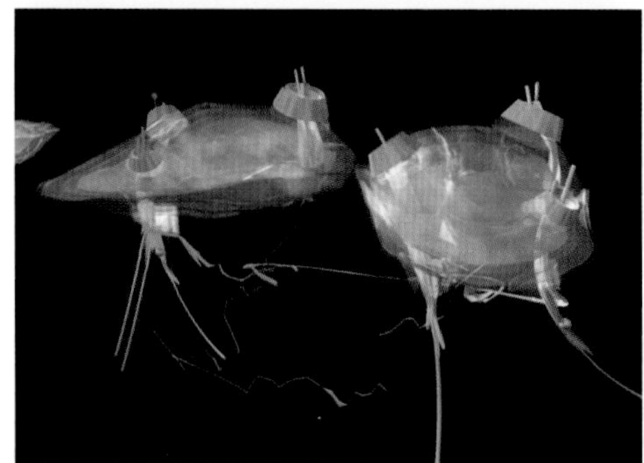

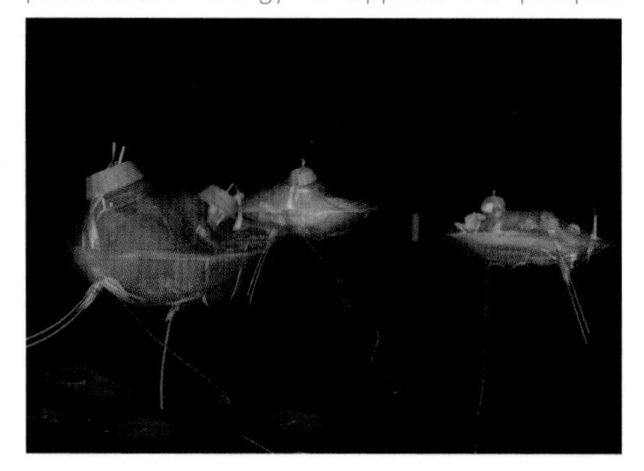

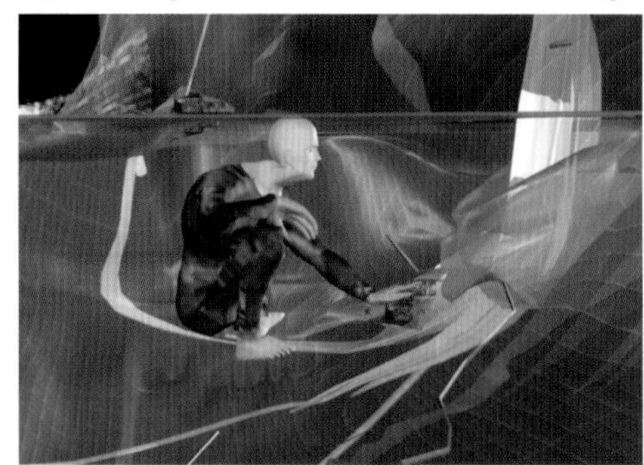

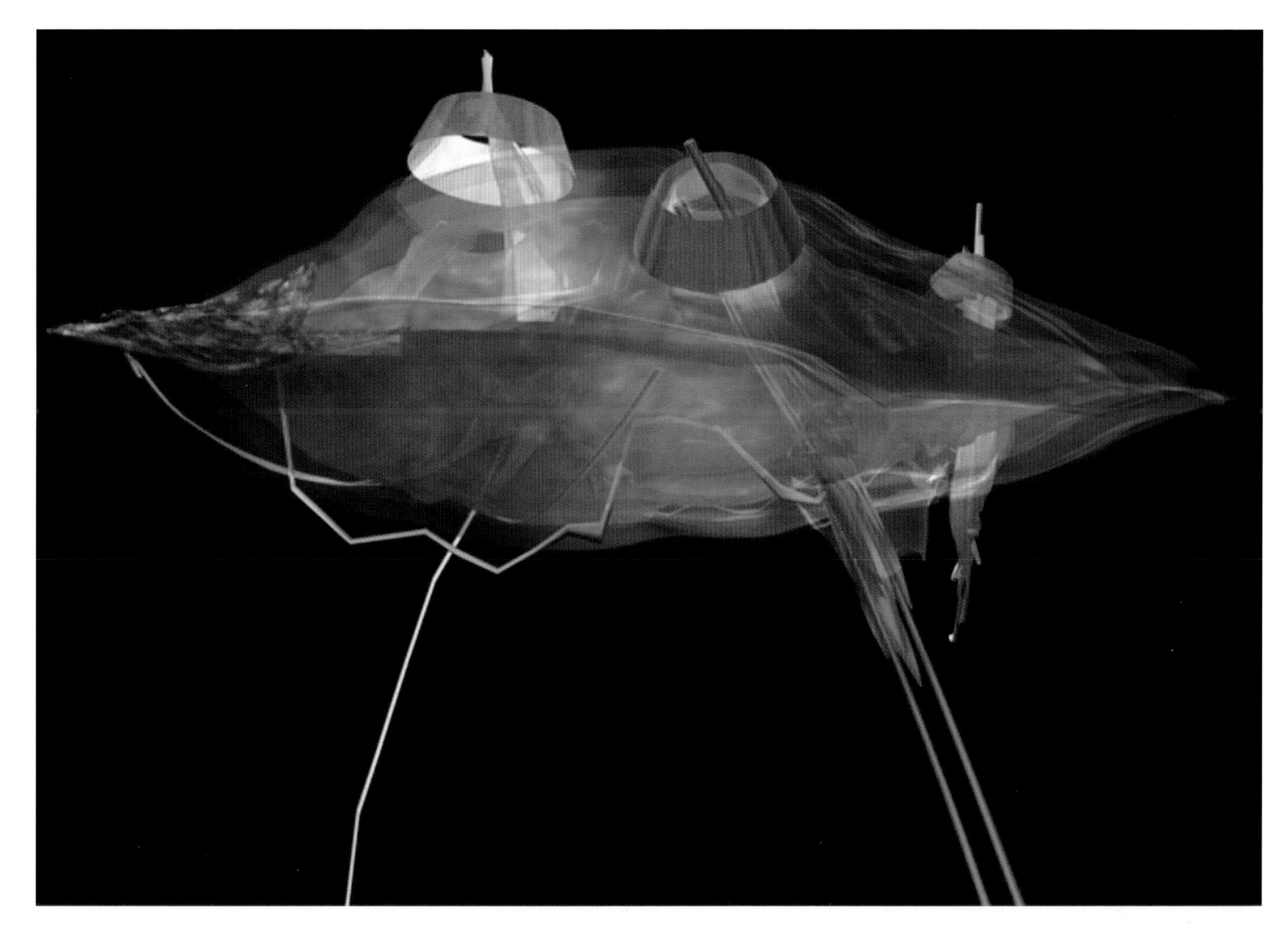

Storehouse

Storehouse is a commission for the 2003 National Design Triennial at the Cooper-Hewitt, National Design Museum. Constructed as a mass-customized modular unit and built from titanium with scrim/fabric-clad wings, Storehouse displays architectural models, drawings, moving images, and books for sale in an intricate system of hard and soft materials. The wings' armature is randomly punctured with shadowbox shelves, creating depth, so that each project displayed is uniquely perceived. The OMD-branded plastic injection-molded base acts both as anchor and seating bench, unfolding, rolling, and adapting to provide a private respite area for museum visitors.

o-md

Prefab ShowHouse

The Prefab ShowHouse exhibits the ideas of prefabrication, flexibility, portability, and compact spaciousness. Located at the heart of Venice Beach's trendy Abbot Kinney Boulevard, the ShowHouse serves as a showroom to display OMD's new work and as a portable model home for inquisitive clients.

More than just being green and modern, the ShowHouse demonstrates OMD's commitment to merging responsible design with new technologies and luxurious details. Its central kitchen/bath core divides and separates the sleeping space from the eating/living space in a compact assemblage of form and function. The steel-frame structure, measuring twelve feet by sixty feet, features a high, sloping twelve-foot-six-inch ceiling and was trucked to its site and set on a temporary foundation.

The prefab design utilizes green technologies—such as tankless water heaters, radiant-heat ceiling panels, and translucent polycarbonate sheets—balanced with high-end amenities like the iPort integrated sound system, Boffi kitchen, and Duravit bathroom fixtures. The ShowHouse offers a space to see the functionality of these elements firsthand, as well as applications of materials such as Kirei board, amber vertical-strand bamboo, and coconut-palm flooring.

Whether briefly situated in an urban lot, momentarily located in the open landscape, or positioned for a more lengthy stay, the ShowHouse accommodates a wide range of needs and functions.